AUSTRALIAN INVENTORS

AUSTRALIAN INVENTORS

Leo Port
with Brian Murray

CASSELL AUSTRALIA

CASSELL AUSTRALIA LIMITED
31 Bridge Road, Stanmore, New South Wales
30 Curzon Street, North Melbourne, Victoria

First published 1978

Edited by Carl Harrison-Ford

Jacket design Kim Falkenmire

Set in 11/13 Times Roman by The Griffin Press, Adelaide

Printed and bound in Australia by Wilke and Company Limited, Clayton, Victoria

F.1078

National Library of Australia
Cataloguing in Publication Data

Port, Leo Weiser, 1922-1978
Australian inventors.

Index.
ISBN 0 7269 6798 x

1. Inventions – Australia. I. Carroll, Brian, 1930-, joint author. II. Title.

608.7'94

Contents

1 Inventive Australians: A story of trial and tribulation 9

2 Ridley and Bull clashed over who invented the stripper-harvester 16

3 Dr Bland's atomic ship 22

4 Geelong publisher James Harrison invented refrigeration, but went bankrupt in the process 26

5 Shearers resisted 'a mechanical contrivance for cutting wool fleece' 32

6 A broken bolt led to Richard Smith's invention of the stump-jump plough 38

7 Louis Brennan's 'ardour for inventions never failed him' 44

8 James Alston improves the power of windmills 47

9 Hugh McKay made a million from the Sunshine Harvester 49

10 Putting the screws on wool 55

11 Experts said George Chaffey's pumping engines would never work 58

12 Lawrence Hargrave walked on water and unlocked the secret of powered flight 61

13 Coolgardie safes cooled goldfields' tucker 68

14 From 'Megaethon' to 'Big Lizzie': Attempts to go where the roads do not 70

15 George Julius turned failure to success with the automatic totalizator 73

16	A German U-boat prodded the British Admiralty into accepting Michell's thrust bearing	77
17	Potter and Delprat put froth and bubble to work on Broken Hill's elusive zinc	82
18	Walter Hume's centrifugal painting machine led to the Hume-spun pipe	87
19	John Dethridge's wheels and weirs	94
20	Lance de Mole's tank was just before its time	96
21	Headlie Taylor taught himself engineering so he could start inventing	98
22	George Nicholas made Australia's first aspirin	104
23	Periscope rifles saved Anzac lives at Gallipoli	107
24	Cliff Howard built a business on his rotary hoe	110
25	Alf Traeger's pedal wireless eased outback isolation	114
26	Falkiner's sugar cane harvester worked well overseas, but not in Australia	117
27	Decades of patient work led to process for making paper from eucalypts	120
28	A vehicle for taking the family to church on Sundays, the pigs to town on Mondays	124
29	Army brass resisted production of the Owen gun, but it became 'the Digger's darling'	126
30	Melbourne pieman asked Uncle Sam for £5 000 000 royalties on his exploding bullet	130
31	From bullet-proof radiators to plywood rafts	133
32	George Shepherd's love of bridge saved the world a lot of pushing and shoving	136
33	Lance Hill put the hoist in rotary clothes lines	139

34 Metcalfe and Wright put some tone in xerography 141
35 Ken Gaunt finally took the backbreak out of cane cutting 144
36 Mervyn Victor Richardson made a million from his mower 148
37 Alan Walsh speeded up chemical analysis with the atomic absorption spectrophotometer 152
38 Frank Cotton's ergometers helped him pick champions 156
39 Vicrail's bogie exchange ended waggon swapping 158
40 David Henshaw gives an old yarn a new twist 160
41 Burt Terry's Rowcar gives new mobility to handicapped children 163
42 Bill Kerruish cuts the work, dirt and danger along with the trees 167
43 Ralph Sarich puts the internal combustion engine into orbit 170
44 Leo Port takes the worry out of catching a lift 174
45 Sirotherm: A 'plum pudding' that makes bad water come good 176
46 Interscan brings the world's aircraft down safely 179
47 'Pure serendipity' in electrical measurements for Mel Thompson and Doug Lampard 183
48 Leo Port's Geometrik clubs make life easier for golfers 186
49 BHP solves the ring-pull problem with the push-button can 187
50 Jack Grant's aircraft slide inflates and floats away 190
51 Edmund Potter shows how TEA can cure smoking 192
52 Ted Pritchard's steam engine can run on alcohol and never pollutes 195
53 Storehouse of a thousand inventions 199
Index 203

Leo Port, engineer, inventor and Lord Mayor of Sydney since 1975, died on 26 August 1978, during the production of this book.

1

Inventive Australians: A story of trial and tribulation

GIVE AN AUSTRALIAN a pair of pliers and a piece of wire and he'll invent something. It may be just to keep his car together, or to prevent his gate from falling apart. It may be of no benefit to anyone but him, yet it will be something that fills a need.

Throughout our history, we Australians have been an innovative lot. We have had to be. Dumped at the opposite end of the world in an often hostile environment, we have had to innovate to survive.

For most of our history we could not send back to Europe for the means of overcoming a peculiar local problem. Even if we had the time to wait for the slow process of communication to work its weary way, we would often have found that those to whom we had exported our problems had no idea of local conditions. The economic success of our major rural industries has been based largely on the inventive work of men who gave us products ideal for the Australian environment; products like the stump-jump plough, shearing machines and irrigation engines.

Yet we have hardly encouraged such people in their efforts to give birth to their inventions, and we have rarely been generous to them after they have succeeded.

Those inventors who found commercial success did so in spite of, rather than because of, the degree of support and encouragement they got from those who stood to gain most from their invention. The success stories of Australian invention belong to men who persisted against every obstacle and every rejection, often to the extent of having to make and market the product themselves.

We find early evidence of that in the shabby treatment meted out to John Bull and John Ridley in South Australia.

In the late 1830s and early 1840s the new settlers there found that their province could grow such bumper crops of wheat that they simply did not have enough labour to harvest it. Nature's bounty seemed to promise only disappointment. Then Bull invented and Ridley built a stripper-harvester that solved the problem, and South Australia became the granary of Australia, a position it long held.

Was South Australia grateful? Not if we are to judge from the refusal of the worthy citizens of the Corn Exchange Committee, who point-blank refused to pay Bull the £40 they had offered for a model or plan of a reaping machine to handle future harvests! Not if we are to judge by the fact that it took 39 years for the South Australian Government to recognize its great debt to Bull! Not if we are to judge by the fact that when it did finally relent, in 1882, it put the value of Bull's invention at a miserable £250! By then the stripper-harvester must have saved South Australia millions.

There was some dispute, it is true, as to whether Bull or Ridley had really invented the stripper. Eventually most people agreed that Bull had been first to conceive the idea, and that Ridley had been first to build a working model. Surely facts that were ascertainable in 1882 should have been more readily so 39 years before.

Fortunately for Ridley, he was independent of the generosity of people he had helped so much. He made his fortune from flour mills and South Australian copper mine shares, and went back to England in 1853. South Australia gave him belated recognition by presenting him with a silver candelabrum in 1861, and by naming in his memory both a scholarship at Roseworthy Agricultural College and a set of memorial gates at Adelaide Showground, erected in 1933, nearly fifty years after he was dead.

Victorian inventor James Harrison could have been wealthy if he had stuck to newspaper publishing. But he had a vision that, with the help of some suitable freezing technique, Australia could get its surplus meat to markets on the other side of the world in saleable condition. Harrison did a great deal of intitial work on the idea, and certainly seems to have invented the world's first domestic refrigerator, but he bankrupted himself in the process. Those who stood to gain most from his success did nothing to help.

Australia's wool industry was just as unhelpful to those who sought to eliminate some of the toil of hand shearing. Higham's original idea for mechanical shearing seems simply to have withered on the vine. Frederick York Wolseley must have been made of sterner stuff than Higham, however, since he persisted with his idea for years until he made a commercial success of it.

But when Wolseley tried to set up his factory in Australia to manufacture an already proven product, he simply could not raise the money he needed. He had spent £15 000 of his own money on his experiments, but he failed in his attempts to float a company in Melbourne, where the get-rich-quick promises of land speculation apparently had more appeal than solidly based enterprises. Wolseley went back to England, set up his factory there, and exported its output to Australia.

Wolseley's shearing machines also ran into trouble with the shearers, who originally went on strike rather than use them. Their attitude may seem bloody-minded, but it has to be seen in the light of the average squatter's reaction to mechanical shearing, which he welcomed not as a means of improving the lot of the average shearer but as a means of strike-breaking. To become proficient with hand operated blade shears was a two-year process, but anybody could shear mechanically after a few days' practice.

While Wolseley was having his trials and tribulations on outback sheep stations and in Melbourne workshops, Richard Smith was suffering for inventing the stump-jump plough, another big breakthrough in the advancement of Australia's rural economy.

Smith was either a 'fool or a lunatic', his peers told him, though by inventing the stump-jump plough, Smith made it possible to farm the mallee lands of South Australia, Victoria and New South Wales. In the three years after it was invented, settlers in South Australia alone took up 500 000 acres of land previously regarded as uneconomical to work.

Smith made nothing at all out of his patents. Yet there was bitter debate in the South Australian Parliament when five years later his well-wishers proposed a grant of £500, a gold medal and a square mile of land at Ardrossan. Smith got his money, his medal and his land, but the bitterness lingered on. The good people of Ardrossan finally provided a memorial for the inventor of the stump-jump plough, whom they identified not as Richard Smith but as his brother Clarence.

This dispute was another in the same mould as the question of whether Bull or Ridley invented the stripper. It highlights just what a long and tortuous process can be involved in getting an invention on the market, no matter how worthwhile it may be.

There are usually at least three stages. Somebody first has to think of the principle behind the invention, the concept that promises to be a breakthrough in solving some real and current problem. Secondly, somebody has to give the concept physical form, to prove that it can do what it is expected to do. This often involves a certain amount of trial and error, of modifying this and modifying that, until everything works, assuming that it eventually does. Thirdly, somebody has to get the product into quantity production and persuade a worthwhile number of people of its merits. At times, one person does all three. At times, it takes three or more people to get the invention through all these stages.

In the case of the stump-jump plough, it seems that Richard Smith conceived the original idea. But he very soon passed out of the picture and his brother Clarence came into it. Clarence was Richard's apprentice, so it was perhaps only natural that Richard should pass over to him the job of bringing his plough into physical being. Yet this act, which probably seemed simple enough at the time, put Clarence into the position where he could be identified, rightly or wrongly, as the person who did most to establish the stump-jump plough as a household word in Australian agriculture.

It is certain that while Richard was lamenting the lack of success of his invention, Clarence was busily getting organized to improve, produce and sell it. From the earliest stages in the development of the stump-jump plough, they seem to have agreed to go their separate ways. Or perhaps they simply disagreed. Clarence set up his plant at Ardrossan, in South Australia, Richard his at Beverley, in Western Australia. Given the nearness of the mallee lands to Ardrossan, it seems obvious that Clarence got the better side of whatever deal there was.

Louis Brennan either showed better judgment or had more luck in the field in which he chose to invent. From the day his torpedo 'excited wonder and approbation' in Hobson's Bay he hardly looked back. Within a few years the British Government had sent for him, paid him £110 000 for his ideas, and installed him in a government job in England. Perhaps it is true, as one obituary said of the Irish-born Brennan, that 'he had all the charm of his race'.

Not that all Australian inventors of machines of war had it so easy. Most of them found it hard to convince the military brass that their ideas had any merit.

In 1905 Melbourne engineer Anthony George Michell patented a thrust bearing that reduced friction greatly, with consequent enormous savings in fuel, particularly in ships, where the new bearing had the added advantage of taking up considerably less space. Michell's bearing waited ten years for the recognition it deserved from the Admiralty, which took it up only after the Royal Navy captured a German U-boat intact and found it equipped with a Michell bearing.

Adelaide engineer Lance de Mole was not even as lucky as that. He proposed the idea of the military tank to the War Office in 1912, only to be told that it was 'not further experimenting with chain rail vehicles'. Four years later, while de Mole's drawings gathered dust in the War Office files, teams of inventors worked desperately to find a way to break the stalemate of trench warfare and came up with something that looked remarkably like de Mole's invention. After the war a Royal Commission declined any financial recognition whatsoever to de Mole.

The same sort of thing happened in World War II. When Evelyn Owen showed his sub-machine gun to an ordnance officer at Victoria Barracks, Sydney, in 1939, he was told the army was not greatly interested in his invention. Sub-machine guns were for gangsters, not soldiers. When within a year or two it became clear that sub-machine guns were very much for soldiers, the Australian Army seemed determined to prevent acceptance of Owen's ideas.

Only when the matter was taken out of the army's hands did the Owen gun get fair trials and the recognition it deserved. But recognition was about all Owen got. For a gun that had saved thousands of Australian lives and saved the taxpayers £2 million in production costs, the Australian Government paid the inventor £5 500 in royalties, and then took back £4 500 in tax. The £1 000 Owen netted from his invention was not enough to cover his expenses.

Some Australians have done well from their inventions. Examples include Hugh McKay, Cliff Howard and Walter Hume and they have one thing in common. They became their own entrepreneurs, sticking with their invention through thick and thin, getting it into production themselves, gradually ironing all the bugs out of it, and eventually building a substantial business on it.

When 19-year-old Hugh McKay demonstrated his combined stripper-harvester-thresher-winnower in 1884, most farmers could see its great benefit. But when McKay approached several Melbourne agricultural machinery manufacturers they all declined to make his machine. Undeterred, the youthful inventor went into business himself and started making the machine that soon became the famous Sunshine Harvester. When the depression of the early 1890s wrecked his infant business, he simply started all over again. He became Australia's largest maker of agricultural machinery. The firm he founded still carries on business at Sunshine, near Melbourne, as Massey-Ferguson (Australia) Ltd. When he died in 1928, McKay was a millionaire.

Cliff Howard's story is almost identical. He was only 16 when in 1912 he first got the idea that some kind of rotary movement could be used to replace traditional ploughing. Howard apparently went straight into business on his own account, without bothering too much about outside help. He stuck with it through the early days of financial strain and it was not until the middle 1930s that his company really got on its feet. Today Howard Rotavators are in use all over the world.

Walter Hume was already in business on his own account when in 1909 he conceived the idea of making concrete irrigation pipes by spinning concrete rapidly in a steel cylinder. He successfully floated a company and exploited his invention until Hume-spun pipes were in use in all parts of the world. Humes Limited is today a leading Australian public company.

At the opposite end of the commercial spectrum from McKay, Howard and Hume was Lawrence Hargrave, the gentleman inventor who almost invented the aeroplane but steadfastly declined to patent any of his developments. Of independent means, Hargrave devoted his life to experimenting with box kites, model aeroplanes and tiny engines. He may have beaten the Wright brothers in being the first man to fly if he had not had such a one-track mind about how such a miracle would eventually be achieved. One of his propeller driven models actually flew, but his dedication to his 'trochoidal theory' led him back to models with flappers.

With the coming of the twentieth century, the pattern of Australian invention had moved away from the farm, shearing shed and blacksmith shop to secondary industry and into the laboratory.

One early example of this change came with the work of such men as Charles Potter and Guillaume Delprat in developing a process for extracting the zinc content of the enormous heaps of tailings that lay like mountains on the outskirts of Broken Hill. They developed the flotation process of mineral separation, which has been described as 'one of the three great advances in metallurgy in the last thousand years'.

Lou Benjamin and John Somerville worked steadily for nearly 15 years in various laboratories until they developed techniques for making paper successfully from Australian eucalypts, something overseas experts had said would never be possible. By the late 1930s their work had made establishment of Australia's modern paper industry possible.

Meanwhile, in the Ford drawing office at Geelong, Lewis Bandt had responded to a simple request from a farmer for a vehicle suitable for taking the family to church on Sunday and the pigs to market on Monday and produced the world's first utility.

These innovators had all enjoyed the support of private enterprise. Firms with an established base and a reliable cash flow had proved less timid than individual investors in putting their money behind Australians who seemed to have an idea worth developing.

Yet there was need for more support for such people and the task of providing it fell to the CSIRO, a body that went through several stages of evolution before it emerged as the Commonwealth Scientific and Industrial Research Organization, an agency of the Australian Government, in 1926.

Until 1939, CSIRO research concentrated mainly on agriculture and processing of agricultural products. Between 1939 and 1945 it gave particular emphasis to food preservation and transport. After 1945 an initial swing back towards agricultural research gave way to expansion of projects geared to manufacturing and processing industries. In the 1960s, rapid development of Australia's mineral resources led to a marked increase in research on mineral processing and techniques of mineral exploration.

Much CSIRO work is necessarily unspectacular and rather anonymous. But it is possible to point to some projects that flowed from steady work by an individual and his team. Dr Alan Walsh, working in CSIRO's Melbourne laboratories, first realized the possibilities of atomic absorption as an analytical technique in the early 1950s. Today atomic absorption spectrophotometers are in use around the world in medicine, pollution control, mineral exploration, and almost every field where rapid and accurate chemical analysis is useful.

Australia's inventors have come a long way since our early days. No longer do they have to struggle on against scepticism and indifference. Ralph Sarich may spend twelve hours a day six or seven days a week working on his revolutionary orbital engine, but he does not have to worry about his financial position. Australia's biggest company, BHP, agreed to stand behind him financially while he developed his engine.

2

Ridley and Bull clashed over who invented the stripper-harvester

---

To the end of their days South Australian farmers John Ridley and John Bull kept up a running argument about who invented the stripper-harvester, the machine that enabled one man to harvest as much grain as five men with scythes or sickles.

Nobody disputed that Ridley was the first man to build an actual stripper that worked. But Bull maintained that Ridley had pirated the basic idea from him. Ridley said his stripper was based on an old Roman harvester, and that he had seen a picture of it in a book.

The story of the Ridley-Bull argument began in South Australia in the early 1840s. Early settlers there found the plains north of Adelaide almost too perfect for growing wheat. Long weeks of sunshine produced crops that called for enormous manpower at harvest time.

A good man with a scythe could cut an acre of wheat a day, and in 1843, South Australia had 43 000 acres to harvest. Soldiers and civilians joined the farmers to get the crop in. Governor Grey even persuaded the New South Wales government to send 150 troops to help with the enormous task.

In an effort to avoid a recurring annual need for such measures, a group of citizens formed the Corn Exchange Committee, and offered a prize of £40 for a model or plan of a reaping machine to handle future harvests.

John Wrathan Bull was one South Australian who had a bumper crop to harvest in 1842. An experienced but not particularly successful farmer, Bull was born in Kent in 1804, and took up dairying at an early age. He migrated to South Australia in 1838 and began dealing in stock, buying sheep and cattle that had been overlanded from New South Wales and selling it to new arrivals.

Like many a South Australian, Bull was forced out of business by the financial crisis of 1839 which almost wrecked the infant colony's economy. He settled at Mount Barker and took up farming as a sideline to his stock dealing.

When his wheat crop was ready in 1843, Bull found it very hard to get labour. But he finally induced five men to come to the farm on the promise of fifteen shillings and a bottle of rum per acre, plus rations, for hand reaping his crop.

Bull would not allow his men to work on Christmas Day, even though the crop was dead ripe, with the heads dropping off under the weight of the plump grain. A fiery hot wind blew all through that day and the next.

John Wrathan Bull.

Bull expected his reapers to start work on Boxing Day, but they did not. He found them in the nearest grog shop, and after some trouble managed to get them to appear for work next morning.

Before the men started harvesting, Bull took the most sober of them to show him that the crop was over-ripe and would need careful handling if much of the grain was not to fall from the heads. He showed him grain already on the ground, then grain actually showing through the gaping chaff on the standing stalks.

To demonstrate how easy it would be to lose the grain, Bull passed his left hand with the fingers spread just below the ears, so the straws passed between his fingers, with the ears close to the palm of his hand. Then he struck the heads

John Ridley.

with a sweep of the edge of his right hand, knocking the clean threshed wheat into the palm of his left.

Bull had for weeks been pondering how a machine could harvest wheat. Now he suddenly saw the answer. He soon tried hitting a bunch of it with a piece of wood, using a circular motion that sent the wheat upwards and across the floor of his barn.

Soon Bull had a rough drawing of a comb and cutters which he showed to his neighbours, none of whom gave him any encouragement. He produced a model of the device to the Corn Exchange Committee in September 1843. It commented favourably on the model but did not part with the prize money.

Whether John Ridley pirated Bull's idea or whether he invented a stripper himself is an argument that will probably never be settled. But Ridley had a full-sized stripper made in October and November 1843, while Bull had only a model.

Ridley was born in Durham, in 1806. His scientific interest showed early, and he built workable electrical machines from the age of seven. At 15 he began to share in the milling business his father had left his mother, and managed it until his mother died too.

In 1839 Ridley arrived in South Australia, bringing with him a flour mill worked by steam. He set it up near Adelaide and ground the first flour from the first South Australian wheat harvest, in 1839. By 1842 he was a prosperous farmer, but he let the farm on a share system so he could concentrate on his milling business.

Like Bull, he gave much thought to the problems of getting in the enormous wheat crop of 1842-3. He devised a stripper and had it built, but did not submit it to the Corn Exchange Committee. However, the Adelaide *Observer* reported on 18 November 1843 that 'a further trial of Mr Ridley's machine has established its success'.

Ridley's stripper had a horizontal projecting comb to grip the stalks, and revolving wooden beaters to knock the heads off the stalk and into a box. The beaters were driven by belts connected to the axles of the main wheels. Horses attached to a pole at the rear pushed the whole machine forward through the crop.

Although various reaping machines had been invented overseas and were in use there with varying success, they were hardly known in Australia. Ridley's

Ridley's 'reaping machine'. In the first models the horses pushed the machine.

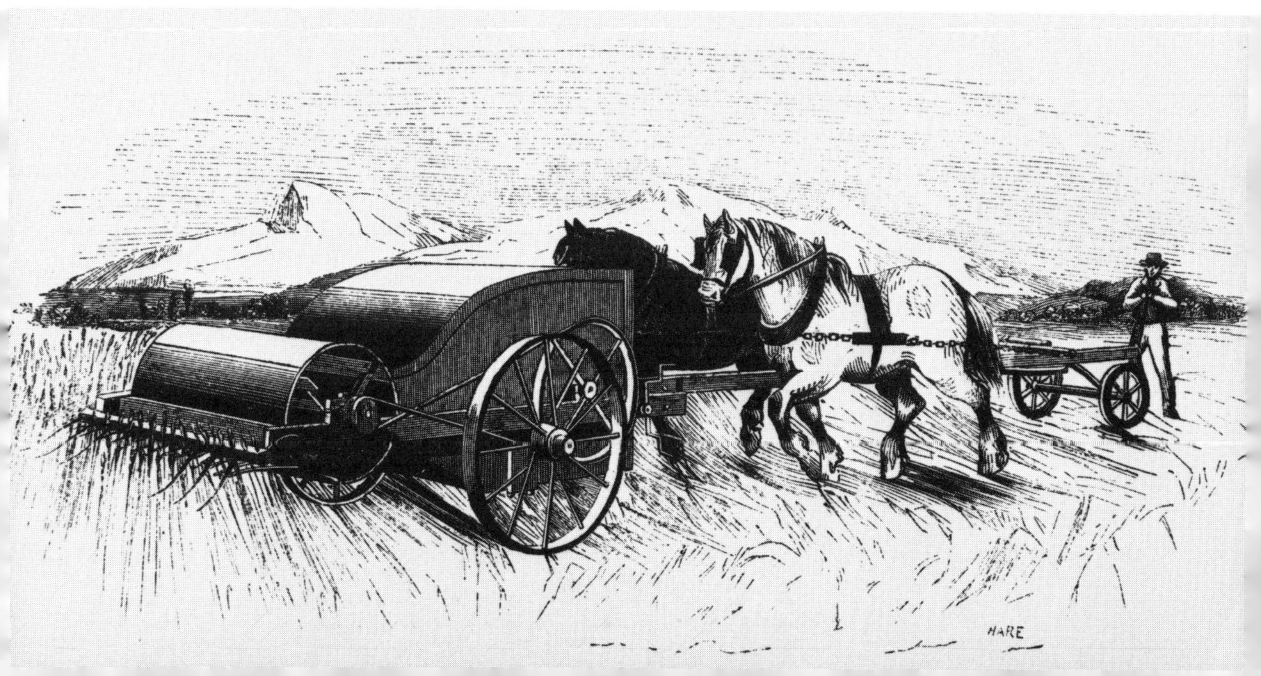

Ridley-type strippers with improved side seating at work near Hammond, South Australia.

machine differed from them in one important aspect. All the overseas machines copied the manual harvesting method of cutting off the stalk at ground level. These were then gathered into bundles, tied into sheaves and arranged in stooks in the field, where they dried out in preparation for threshing. The South Australian machine cut the stalk just below the grain heads, threshed the grain from the heads, and fed it into a box.

Ridley demonstrated the machine on his own 70 acre farm, one man guiding the two horses, the other the machine. They harvested the 70 acres in seven days, five acres per man per day; five times the rate of manual harvesting.

One observer wrote in 1844: 'I have seen two hundred acres reaped and threshed by the [Ridley] stripper and the proprietor, an old colonist of New South Wales, thinks highly of it. Its charge was ten shillings an acre . . . ten acres a day was its performance.

'It consists of a rake which catches up the ears to a width of about eight feet, a cutter which revolves and takes them off the rake, and a chamber posteriorly which collects the grain from the cutter and is emptied by a trap-door at the end of every lane on a canvas laid on the ground, whence the grain is shovelled up into bags and requires nothing subsequently but winnowing. The machine both reaps and thrashes.

'The drier the corn and the riper, the better the machine works. It is peculiarly adapted for this hot climate where grain ripens to a danger of falling in reaping.'

In 1847, a writer in the *South Australian* reported that 'the reaping machine invented by Mr Ridley is now almost universally adopted in this country. One machine maker has no less than seven in progress and the number in the colony is upward of one hundred.'

Adelaide manufacturers were busy making the strippers and there was a good export trade in them. One manufacturer, Adamson of Mount Barker, arranged the stripper so the horses could pull it from the front instead of push it from the rear as Ridley had designed it.

Ridley declined to patent his stripper, and refused all suggestions of reward. He was reasonably affluent, thanks to the success of his flour mills and his interests in copper mining. He went back to England with his family in 1853, travelled for some years in Europe, and then spent his time on minor inventions. He tried to adapt the stripper to English conditions by adding a cutting knife under the comb, and in 1862 he entered three of his machines in the London Exhibition. He also exhibited there a cinder-sifter and a straw cutter he had invented. In 1868 a Leeds firm made 12 of the London Exhibition style of reaping machine and sold them in South America.

Ridley spent much time and money in his later years distributing literature on temperance and religious questions. Although he had not sought to make money from the invention of the stripper, he was nevertheless concerned about Bull's claims that the true credit for inventing the stripper belonged to him.

He lived to see the South Australian Government acknowledge that Bull was entitled to at least part of the credit, when in 1882 it awarded him £250. This was 39 years after his work on the 'locomotive thresher'.

Bull joined the gold rush to Victoria in 1852, had some success there, and went back to South Australia in 1853. He planted a vineyard and traded in his own and other colonial wines. His book *Early Experiences of Life in South Australia* was published in 1878, eight years before his death.

Did Bull or Ridley really invent the stripper? Or did they both hit upon the idea about the same time? Nobody can really say. But we can be sure that they had started a revolution in grain harvesting methods all around the world.

3

Dr Bland's atmotic ship

MEN FOR MILLENIA have wanted to fly and a man with so active a mind as Sydney surgeon William Bland was bound to give some thought to such a challenge. His interest in flying was intensified by another of his interests, the problem of transport and communication between Australia and the rest of what white men then saw as the civilized world.

Born in London in 1789, Bland trained for the medical profession and in 1809 became a surgeon in the Royal Navy. He was serving in the East Indies when he quarrelled with the purser and in the ensuing duel shot him dead. He was tried for murder and in 1813 transported to Australia for seven years.

Luckily he arrived in Sydney while Lachlan Macquarie was Governor. Macquarie was inclined to be soft on convicts who were also professional men, particularly if they could be useful to him. Bland was put in charge of the mental asylum at Castle Hill and pardoned in 1815, after which he began practice as a doctor.

Bland became an active politician as well as a doctor, and applied his inventive mind to solving a problem that was bothering Australia's major industry, that of spontaneous combustion of the cargo in wool ships. Bland published a pamphlet on the subject in 1843, describing a device for circulating carbon dioxide throughout the cargo. This deprived any latent fire of its oxygen supply. Bland exhibited a model of his device at the International Exhibition in London in 1851.

By then Bland had a more ambitious project in mind.

'Aerial navigation has long since appeared to me beyond doubt practicable, and if so is calculated to confer many highly important benefits on man,' he wrote in the Sydney paper, the *Empire*, in 1860, describing the airship he had first proposed nearly ten years before.

Bland claimed credit for 'the invention of a vessel with which it is proposed to navigate the atmotic regions, and that too with all the facility with which they are traversed by their own proper denizens'. Bland's word 'atmotic' derived from the Greek word atmos, meaning vapour.

Bland published drawings of a long, powered balloon that looked remarkably like the Zeppelins of the early twentieth century. He proposed a float (as he called the balloon) between 60 and 150 metres long and between eight and 15 metres wide. He calculated its total lift would be about five tonnes. Since a car and steam engine would weigh between three and four tonnes, this left a payload of between

Dr William Bland.

one and two tonnes, adequate for passengers, mail, and such precious cargo as gold and gems.

Bland pointed out that the buoyancy of the ship could be increased indefinitely merely by increasing the size of the balloon. In view of its elongated shape, size would not impede its ability to pass through the air.

Bland speculated that his atmotic ship would travel at 50 knots. It all depended, he said, 'on the various degrees of velocity and the various directions of the several successive currents or strata of which the atmosphere may consist at different heights and in different parts of the globe'. He thought the average

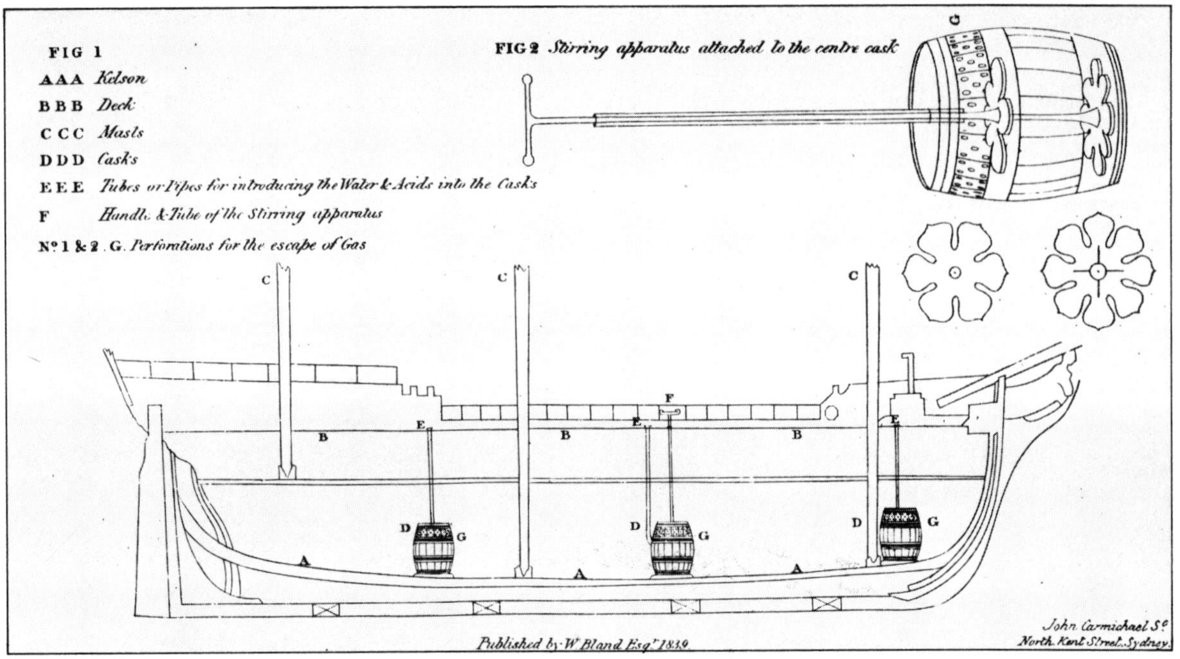

Bland's apparatus for overcoming spontaneous combustion in wool ships. The barrels were always charged with quicklime. At the first sign of fire the crew poured water and sulphuric acid down the pipe. The handle was for stirring the mixture, which produced carbon dioxide to smother the fire.

speed of these would be 40 knots, to which he added the 10 knots that would be provided by the propellers. These would be driven by the steam engine carried in the gondola slung under the float.

By taking a direct route from Sydney to London, the atmotic ship should be able to cut the distance from 12 000 to 6 000 miles, Bland argued. At the pro-

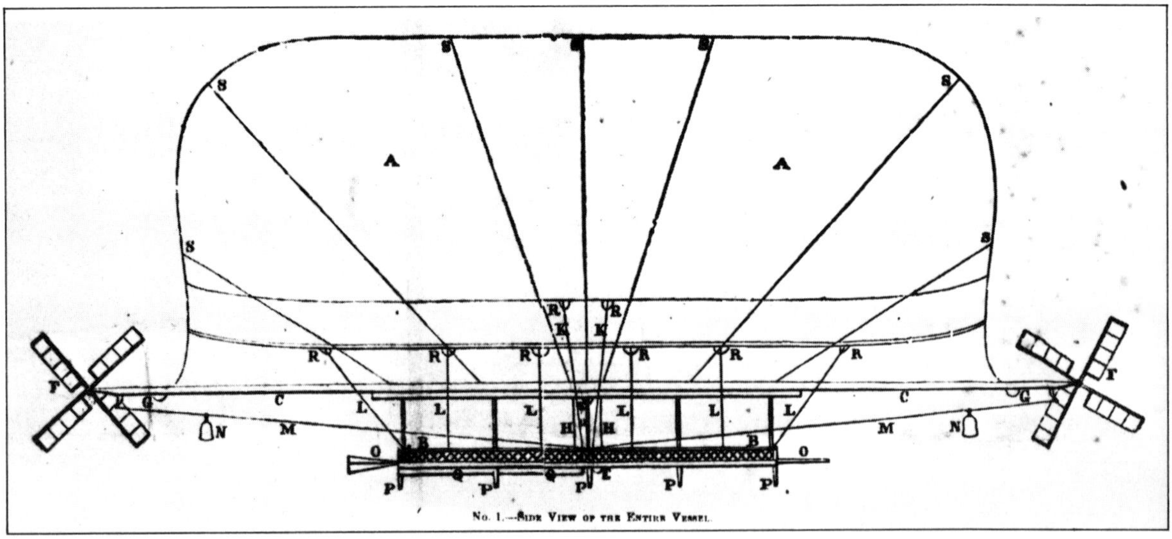

Dr Bland's atmotic ship. The gas-filled balloon was to support the iron car, which would carry fuel, boiler and steam engine, plus passengers and freight.

jected speed of 50 knots it would therefore cut the time for the journey from two or three months to four or five days.

To some people, aerial navigation might seem to have its risks, Bland agreed. For one thing, atmotic ships might fall into the wrong hands. 'Could this method of travelling tend to place the different races of man too much on a par with each other as to their agressive powers?' he asked. Bland apparently foresaw the possibility of aerial bombing.

But Bland disposed of possible technical objections to his atmotic ship.

Some people might think it at risk from the electric fluid in the atmosphere, particularly because of the quantity of iron that would be needed for its steam apparatus. This was negated, he said, by actual experience. Any danger from the concentration of iron would be considerably less than that incurred by an iron ship when it was weatherbound.

Another possible risk was from the proximity of the fire in the steam generating appartus to the highly flammable contents of the gasbag. Bland felt the six metres between the two eliminated this danger! Anyway, the chimney would have a spark arrester.

The risk that the float would burst through 'a diminution of the requisite external pressure in consequence of the greater rarity of the air in the higher regions' did not bother Bland. The usual net would be thrown over the float, which would be divided into a 'due number of more-or-less completely separate compartments'. And there would be escape valves to let the gas out as necessary.

Bland's ship would be exempt from the kind of accident usually associated with ordinary balloons, which had a tendency to come to grief during either take-off or landing. He said such mishaps had been 'amply provided against', but did not go into details.

Bland did not live to see powered flight become a reality. He died in Sydney in 1868.

4

Geelong publisher James Harrison invented refrigeration, but went bankrupt in the process

IMAGINE A WORLD without refrigeration. Australia's meat and dairying export industries simply would not be possible. Air conditioning would be unknown. And we would all have very different domestic and eating arrangements. Almost every home in Australia now has its own refrigerator, protecting our health, reducing the number of our shopping trips, and bringing us the produce of the world near its flavoursome best.

But the man who invented refrigeration, Geelong publisher James Harrison, went entirely unrewarded. In fact, the cost of his invention bankrupted him.

The son of a Scottish fisherman, Harrison was born in 1816. Apprenticed to a printer in Glasgow, he soon moved to London, then to Sydney, and on to Melbourne, where he arrived in 1839. He worked with the fiery John Pascoe Fawkner, owner of the *Port Phillip Patriot*, who sent him with an old press and some discarded type to start a newspaper in Geelong.

By the early 1840s, Harrison was sole owner of the *Geelong Advertiser* and doing quite well. The discovery of gold near Ballarat in 1851 brought great prosperity to Geelong, in which he shared.

Harrison might have continued to prosper as a publisher except that he had already developed a passion for inventing. Using ether in his printing works to clean type, he noticed that as it did its intended work it also made the type quite chilly. Harrison soon realized the potential of this phenomenon for solving one of Australia's great economic problems.

As Australia's pastoral industry had developed, many people had begun to wonder about ways of preserving her surplus meat so it could be exported to markets on the other side of the world. By 1847 the grazier brothers Henry and William Dungar had started canning meat at Newcastle, but canned meat in those days was used mainly by explorers and on ships. To exploit the major markets, Australian exporters would need some better method. In the meantime, the only way they could find an export market for surplus livestock was to boil it down and ship the tallow.

Harrison was very conscious of these problems and decided to experiment with ether to see if he could make artificial ice. Until then, all ice in use for preserving food was natural, dug from ponds in cold climates and transported to all parts of the world. Such ice as was available in Australia had usually made the long voyage from the north-east of the United States in sailing ships via Cape

Geelong publisher James Harrison.

Horn. Such an expensive process as this offered no hope of solving the problems of preserving Australian meat in export quantities.

Harrison had soon developed his ideas to the point where he had a foot pump made to circulate ether through a coil. The ether was first throttled down by a cock, so that it would expand suddenly into the coil. In this, Harrison was using the well known principle that a gas cools rapidly if it expands. This is exactly what had happened when he was cleaning his type, for evaporation of a gas is really expansion.

In a further development of his idea, Harrison had the coil immersed in a barrel of water, and succeeded in making his first artificial ice. He began an extensive correspondence with several eminent scientific figures in Britain, including the famous Michael Faraday who has often been called the 'Father of Electricity'.

Harrison set up his equipment in a cave in Geelong. On one occasion he was badly injured when it exploded. The trouble with ether as a refrigerant is the fact that it is highly flammable.

By 1857 Harrison had a small factory on the banks of the Barwon River at Geelong, and was turning out three tonnes of ice a day. He experimented with methods of freezing meat, fish and poultry. To publicize his invention he often gave visitors a whole fish or a slab of meat frozen inside a block of ice. The Bendigo brewers, Glasgow & Co., adopted Harrison's principles in a pioneer mechanical refrigerator.

Harrison's ice was used on more than one occasion to reduce the fevers of sick people. A report in the *Geelong Advertiser* of 8 September 1858 stated: 'The ice making machinery has twice during the past week been utilized in providing ice in urgent cases of sickness. The ice has been produced in blocks of seven pounds in about two hours after receipt of the order. In each case, it is said, the prompt application of the ice has been the means of saving life.' As publisher of the *Geelong Advertiser*, Harrison understood the power of the press in promoting a cause.

In 1857 Harrison went to England and Europe to study developments in refrigeration there, and to have a larger compressor made. He took a scale model of it to the Continent, where many people were impressed but declined to pay the price Harrison asked for the use of his patents. In 1859 he had his new plant loaded on the ship *Tricolor* and brought it back to Geelong.

Harrison set up a new plant at Western Beach, Geelong, and hoped that money would soon start flowing in from the sale of ice. But it did not. The plant turned out far more ice than Geelong could possibly use, and most people still preferred the 'natural' ice from America.

Although Harrison was still proprietor of the *Geelong Advertiser*, much of the daily control of it had passed to his younger brother, Daniel. In 1860, the paper was involved in the heavy expense of moving to new and larger buildings and the financial drain of this, added to the ice making experiments, forced Harrison into bankruptcy. He had assets of £22 000 and debts of only £14 000, but he was forced to sell the paper to pay his debts. The new owner, Geelong businessman Alfred Douglass, had enough respect for Harrison's talents, however, to keep him on as editor.

In an attempt to find a market for his new plant, Harrison moved it to premises in Franklin Street, Melbourne, where it went on turning out ten tonnes

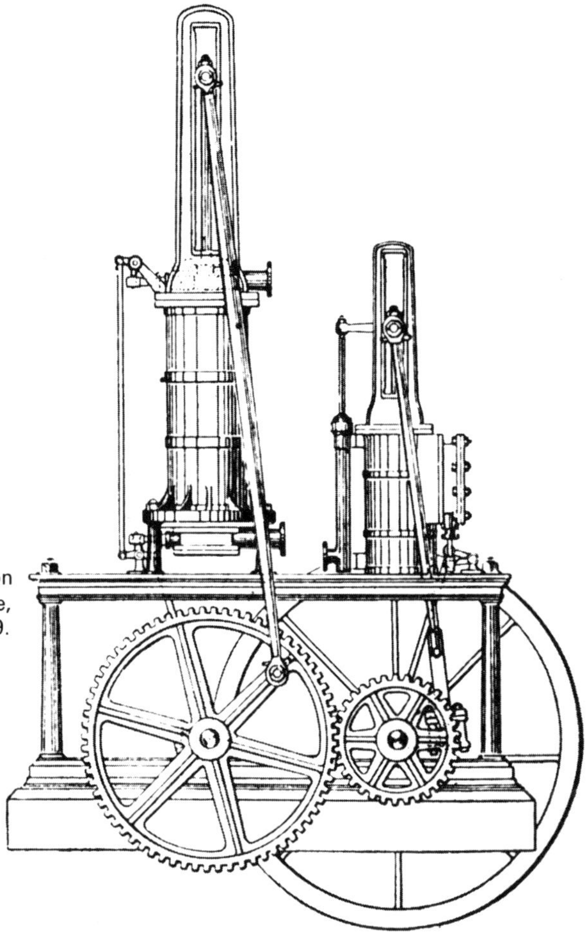

Harrison's vapour compression refrigeration machine with steam engine, 1859.

of ice a day. But he still could not sell it. People still preferred 'natural' ice, and importers advertised it heavily. In Melbourne, Harrison became editor of the *Age*.

He kept up his faith in the future of freezing meat and other foods and at the International Exhibition in Melbourne in 1873 he showed a box of frozen foodstuffs packed as an export consignment. It remained under supervision at the exhibition for 97 days, and when it was opened the contents were found to be perfectly fresh and sweet. Harrison was awarded a gold medal.

With the assistance of Sir James McBain, who contributed £2 500 towards the expenses, Harrison now prepared for a full scale shipment of meat to England by the sailing ship *Norfolk*. Unfortunately the cargo did not reach its destination in a marketable condition, due to a combination of factors: rough handling during loading, restrictions on operation of the refrigerating plant because of

The freezing room on board the steamship *Strathleven*, 1879.

fire insurance rules, the rough and protracted voyage, and the understandable inexperience of the crew. Harrison was ruined financially.

On the failure of this venture, Harrison went back to London and settled down to journalism. He came back to Australia in 1893, aged 78 and still inventing. He hoped to set up a plant near Geelong to extract soda from limestone. He died the following year, half-way through dictating an *Age* article to his daughter.

Harrison had some success with smaller inventions. The Geelong and Western District Horticultural Society awarded him a silver medal in 1852 for a 'self-acting beer cooler', details of which have been lost. He also devised some improvements in quartz crushing and evolved an amalgamating process to help miners recover more gold. And a company in Scotland used one of his refrigerators to distil paraffin.

Harrison's work was carried on by French-born Eugene Nicolle. The two men had been associated in the work of the Sydney Ice Company, which Harrison formed with P. N. Russell in 1860 but soon sold. Nicolle in 1861 took out a patent on a method to liquefy ammonia, which meant he could avoid the hazards of the highly flammable ether. In 1863 the Sydney Ice Company began to use Nicolle's machine at its Darlinghurst plant.

In 1865 Nicolle met Augustus Morris, a New South Wales grazier who had long been interested in the problem of preserving Australia's surplus meat. He had tried unsuccessfully in 1857 to raise £100 000 to be given to anybody who could invent a commercial process for freezing meat and exporting it overseas.

Morris in 1866 persuaded Sydney trader Thomas Sutcliffe Mort to fund Nicolle's continued experiments. Morris went to England to study meat preserving methods and to patent a machine invented by Nicolle.

By 1872 Mort and Nicolle had a small experimental plant in operation at the back of the Royal Hotel in George Street, Sydney, and in 1873 they set up works at Lithgow and Darling Harbour to do refrigeration on a commercial scale. They killed and froze meat at Lithgow, then brought it by train to Sydney for sale from their freezing works there.

Mort spent £100 000 on experiments in refrigeration, but lost it all. Early in 1877 he chartered space on the ship *Northam* and had refrigeration machinery installed, but due to a defect in it the ship sailed without its cargo of frozen meat. Mort died before the first shipment of frozen meat left Australia in the *Strathleven* in 1879, using a British chilled air process, but Harrison was still alive.

Nobody seems to know what happened to Harrison's original ice-making machine, but the small steam engine which almost certainly is the one used to drive it turned up on a farm at Heathcote, Victoria, in 1970. It was in 'near mint' condition, and its then owner, Heathcote bee-keeper Robert Campbell, was using it to power a honey extractor.

Being a member of the Lake Goldsmith Steam Preservation Society, Campbell had kept the engine in good condition. Once he realized its historical significance, however, he retired it from service. He was able to trace the history of the engine via Tait's boathouse at Eastern Beach, Geelong, where it was used to haul boats up the boathouse ramp.

After the boathouse was pulled down the engine went to a garage at Barwon Heads, where the boiler was used to heat water for household use and the engine to drive a lathe, grindstone, and one or two other pieces of equipment in the garage. When the owner died, the engine and boiler found their way to a second-hand machinery shop run by his sons in Geelong, and finally to the Heathcote farm.

In 1967, the Australian Institute of Refrigeration described James Harrison '(in spite of works of reference favouring T. S. Mort) as unquestionably the authentic practical inventor of commercial refrigeration'. His tombstone, paid for by public subscription in 1896, said with ironic truth 'one soweth, another reapeth'.

5

Shearers resisted 'a mechanical contrivance for cutting wool fleece'

AUSTRALIA'S ECONOMIC DEPENDENCE on wool in the nineteenth century meant a life of drudgery for the army of shearers who moved from station to station to part the wool clip from the bleating jumbucks. Added to the pain of constant stooping was the need to open and close the shears by hand all day every day.

You'd expect then that Australia's shearers would have welcomed the invention of some means of easing the task. But most of them stubbornly resisted the development of shearing machinery. Some feared it would make them redundant, others simply that the comparative ease of mechanical shearing was a threat to their image as skilled workers. Squatters soon made it clear, too, that they would use mechanical shearing for strike breaking. It took two years to train a hand shearer, but anybody could shear mechanically after a couple of days' practice.

The first sheep shearing machines came soon after the overseas invention of a horse clipping machine. Many people felt that shearing sheep must be similar to clipping horses, but they soon found that cutting wool was very different from cutting horse-hair. Still, a few people persisted with the idea, and in 1868 Melbourne compositor James Andrew Bolton Higham patented a 'mechanical contrivance for cutting wool fleece'. It was like 'a large hairbrush, the mechanism taking the place of the bristles, leaving the handle for the grip'. The two-bladed cutter revolved on a comb plate. Higham, however, was apparently unable to make a commercial success of his invention.

J. A. E. Gwynne in 1870 patented a sheep shearing machine similar to that produced by Moore and Newman in England the year before. Gwynne's machine was rather too complicated for commercial success, but it did establish the principle of a flexible link between the shear and the shaft, an important development. It also introduced a spring in the handpiece to provide tension, and bevel gears to reduce noise.

Meanwhile, former jackeroo Frederick York Wolseley, later to be famous as the founder of the Wolseley car firm, had been working on the problem of how to shear mechanically.

Wolseley, who was born in Ireland in 1837, arrived in Melbourne in 1854 and went to work as a jackeroo on Thule sheep station on the Murray River. After five years there he acquired an interest in the station, and also in Cobram station. By 1871 he owned Toolong in the Murrumbidgee district, and by then he must have been well aware of the problems of shearing.

Frederick York Wolseley made mechanical shearing a practical proposition.

Financed by his brother Garnet, the famous Field Marshal and Viscount, Wolseley in 1868 began experiments on a machine for shearing sheep, and in 1872 he produced a working model in Robert Savage's workshop in Bourke Street, Melbourne. This machine failed when tested at Euroka Station, near Walgett, on the Darling River, in north-west New South Wales.

Wolseley went to England in 1873 and studied horse clippers made in Birmingham by Clarke and Company. He and Savage persisted with years of

Blade shearing, the old way. Shearers in South Australia, late 1880s.

experimental work and trials at Euroka, which Wolseley bought in 1876. On 28 March 1877 they patented the first commercially successful shearing machine in the world. It was driven by horse power. George Gray, the man who hand-forged the parts, was probably the first man in the world to remove a complete fleece mechanically.

Wolseley and Savage took out a second patent in December 1877, but their machine still had serious problems with the drive mechanism, and imposed physical limitations on the shearer's movements. Further experiments gradually eliminated most of the problems, and on 13 December 1884 Wolseley and Robert Park, another of his collaborators, patented an 'Improved Shearing Apparatus' which included a cog-gear universal joint.

Wolseley enlisted the aid of a former employee John Howard, of Clark and Company, who had begun making horse clipping machinery in Melbourne. Wolseley bought out Howard's patents and engaged him as a mechanic at Euroka at £3 a week. He and Gray worked together, and made several improved machines including one with a drive in which friction from a large wheel on an overhead shaft forced a small cone to rotate on the upper end of a flexible core. This worked so well during the 1885 season that Wolseley went to Melbourne to form a manufacturing company and arrange public demonstrations.

The obvious way to demonstrate the shearing machine was to pit it against the old hand method. So early in 1887 Wolseley took his two best shearers, Hassan Ali, a Sudanese, and Jack Joy, to Melbourne to compete against champion shearer Dave Brown. In a match to shear three sheep, Brown with hand shears finished before Ali, and the judges declared the sheep well shorn. But then Ali ran the machine shears over the sheep and cut another kilogram of wool from them.

After he had made some further improvement to the handpiece, suggested by William Ryley, Wolseley set out on a series of demonstrations all over eastern Australia and New Zealand. At Normanby station in Queensland, a match between six machine shearers and six hand shearers produced a convincing win for the machines. Dave Brown and another blade shearing champion, Charles Shepherd, declared themselves in favour of machine shearing.

Wolseley's big break came in 1888, at pastoral magnate Samuel McCaughey's famous Dunlop station, at Lough, on the Darling River near Bourke. James Wilson, manager of the station, had seen one of the Melbourne demonstrations and was convinced of the merits of machine shearing. He had a full stand of 40 machines installed for the 1888 season.

Sheep shearing with machine shears, 1896

Machine shearing, the new way.

But Wilson had not allowed for the reaction of the shearers. When 40 of them were selected and offered the jobs as machine shearers, they refused to sign on. They objected to a clause in their contract saying they must supply their own combs and cutters, but they clearly were refusing to shear with machine shears. They walked off the property, set up camp on the opposite side of the Darling, and picketed the station. Wilson had 184 000 sheep to be shorn and 40 Wolseley shearing machines standing by, but nobody to operate them.

Two of Wolseley's men, Howard and Shepherd, who had installed the machines, crossed the river each day and mingled with the strikers, never missing a chance to tell them how machine shearing would help shearers. Howard finally won them over when he accepted a challenge to swim the flooded Darling. This appealed to the sporting nature of the strikers, who agreed next day to use the mechanical shears. Wilson agreed to supply the disputed combs and cutters, and the strike was settled.

Trouble continued, however, with some of the shearers deliberately breaking combs, cutting the main pulley belt driving the machines, and turning in miserable tallies like 50 sheep a day. But gradually the shearers came round, and Dunlop became the first property in the world where the entire wool clip was taken off mechanically.

One of the blade champions won over to machine shearing at Dunlop that year was Jim Davidson. He became one of Australia's leading advocates of machine shearing, demonstrating it all over Australia, working in factories that made it, installing it in shearing sheds, and talking shearers into using it.

In spite of the conversion of the reluctant shearers at Louth in 1888, there was still considerable resistance to the idea. When Jim Davidson and John Howard went to Walgett show in 1890 to demonstrate their machines, a local publican was so hostile that he contrived to have them arrested and gaoled in the hope of keeping them locked up until the show was over. Fortunately, a local magistrate ordered their release just in time for them to rush to the showground and show their wares while there was still an audience. The next day, however, the diminutive Howard was beaten up by the burly publican.

Davidson overcame all sorts of obstacles. When heavy rain prevented him getting a steam engine into place to drive his shears for a demonstration at Hughenden, Queensland, he had a heavy bullock waggon jacked up and an iron bar fixed to its back wheel, around which he ran the belt that drove the machinery. Volunteers kept the wheel turning all day while Davidson showed his shears.

Wolseley owed at least part of his commercial success to Herbert Austin, who in 1884 at the age of 18 left his native England to serve his apprenticeship in his uncle's small engineering workshop in Melbourne. Austin worked with several Melbourne firms and then became foreman in Robert Park's workshop, where Wolseley's equipment was being made.

Austin made several helpful suggestions to Wolseley, and the two men co-operated for some years in the development of shearing machinery. But Wolseley could not raise the money to start large scale manufacture in Australia so he went back to England in 1889, set up his own plant in Birmingham to make his machinery, and in 1893 brought Austin back as production manager. Wolseley's health failed and Austin took over completely in 1894. A year later the Birmingham plant turned out the first Wolseley car.

Austin stayed with the company after Wolseley died in 1899, but he left in 1905 to set up his own plant and produce Austin cars. He spent the rest of his life, until 1941, expanding his company. By then mechanical shearing had long since replaced hand shearing throughout the world.

6

A broken bolt led to Richard Smith's invention of the stump-jump plough

MOST FARMERS WOULD curse when a bolt broke in their plough. But South Australian engineer-turned-farmer Richard Bowyer Smith soon noticed that his plough worked better with a broken bolt than it had ever worked before.

Smith was ploughing some stumpy ground one day in the 1870s when one of the bolts holding the arm carrying the mouldboard broke. Smith carried on ploughing, wondering no doubt how long it would be before the damaged plough gave up completely. But now, instead of the share crashing heavily into every concealed stump, it tended to ride over them and return to its work.

Smith soon realized that his broken plough had within it the germ of an idea that would overcome a major problem of Australian, particularly South Australian, farming. In the mallee country to the north-west of Adelaide, clearing the light timber above the ground was not a serious problem to a determined settler. But grubbing out the mallee roots was something else. Land that cost a pound an acre for the freehold might cost anything up to seven times as much to clear and grub. Yet until the ground was grubbed, ploughing by normal means was hazardous. Every time the plough struck an underground obstruction there was a definite risk of injured horses or a broken share.

One enterprising farmer named Mullens overcame this problem by inventing a process that became known as 'Mullenising'. He drove spikes into the underside of a V-shaped log and had his horse drag it about his cleared but ungrubbed land over which he had previously broadcast his seed. The Mulleniser effectively buried the seed, and Mullens soon had a good crop of grain. He was even able to harvest it with a mechanical stripper. Other farmers were soon using the same idea. But Mullenising was a poor substitute for ploughing.

Once the idea of the stump-jump plough had occurred to Richard Smith, he soon got to work with his brother Clarence, who was his apprentice, and put together a prototype. The share, instead of being rigidly attached to the frame, was suspended by a single bolt. When it met an obstruction it merely tipped back and rode over it. Once the share was over the stump or rock it was brought back into the ploughing position by a heavy weight attached to it at the end of a lever.

Richard Bowyer Smith in 1917 when he was 81.

The Smiths went on to produce a three furrow stump-jump plough. But it was too heavy and cumbersome for a team of horses to move, and when a team of bullocks was hitched to it they bolted, smashing the plough into some nearby trees. The two brothers persisted however, and in August 1876 they successfully demonstrated a working prototype of a three furrow stump-jump plough to some farmers.

Later that year, the Smiths exhibited a one furrow stump-jump plough at Moonta show, on the Yorke Peninsula, where it won first prize. The *Farmers' Weekly Messenger* said that the plough would, 'if adopted by agriculturalists, cause a complete revolution in tilling uncleared land'. It would enable farmers opening up new lands 'to till the soil at once, and have the proceeds of their crops to assist them in clearing it as time or funds allow'.

But some of the local farmers were not so enthusiastic. One told Richard Smith he was 'either a fool or a lunatic . . . A plough's got to go in hard ground and stay in unless you only want to tickle the soil.' Another thought the new invention might 'be alright for a slovenly farmer'. 'Fancy seeding while stumps are still in the ground,' he said. 'I believe in getting the land clear first, even if it does take a few years.'

Some even judged Richard Smith's invention by his threadbare clothes. He had put all his money into the stump-jump plough, and had none left to buy a new suit for the show.

The first 'stump-jumping plough', the Vixen,
a three-furrow version made by the Smith Brothers in 1876.

Smith took out a provisional patent on his invention on 19 February 1877, which was valid for one year. But his early ploughs did not perform well, and there was little demand for them. When the patent expired Smith could not afford to renew it.

But one farmer was convinced of the merits of the stump-jump plough. Macklin, who had bought the original three furrow 'Vixen' plough wrote to

Clarence H. Smith's implement works at Ardrossan, South Australia, 1906.

An early 1900s advertisement for the Sun-Twin disc stump-jump cultivator, which 'Works Wonders in Stumpy and Stony Paddocks'.

Richard Smith: 'I have ploughed about 25 acres with your Vixen plough and am glad to say it has done its work admirably well. It never fails to rise over or slide by a stump when it comes in contact with one, and enters the ground immediately the obstruction is past.'

Clarence Smith took over the task of perfecting the stump-jump plough and spent four years steadily eliminating its faults. In 1880, satisfied that he had at last built a successful plough, he built a small workshop at Ardrossan, and went into the business of making them. He was then 25. His Adelaide material suppliers, Harris Scarfe & Company, helped the venture by granting him a year's credit.

By allowing his original provisional patent to expire, Richard Smith had created a situation where anybody could make and sell stump-jump ploughs. Once the plough was a commercial success, there was no shortage of people willing to do just that.

The Speaker of the House of Assembly in South Australia, Robert Ross, asked the government to make a grant to Richard Smith in recognition of his work. Enthusiasts for the stump-jump plough claimed it was the major influence in the settlement of over 500 000 acres in South Australia between 1878 and 1881.

In one of his letters to Ross, Smith gave a picture of the tribulations that can face a struggling inventor: 'I hope you will excuse the shabby appearance of my certificate, but I and my plough and all that pertain to it have been huffed, cuffed and kicked so much for the last nine years that at last I found myself in a hut in which there was not room for my papers, so I had to stow them in a box outside, and one rainy night a pig capsized them and made a bed of them.

'My invention has cost me some money, some anxiety, and condemned my little ones to all the miseries of poverty and banishment in the bush, whereas if I had been a successful cricketer, a good bowler or a rifle shooter without pluck, a Blondin or an acrobat, I and mine would have escaped all these ills.'

Once there was talk of money for the inventor of the stump-jump plough, various other people soon claimed the credit rightly belonged not to Richard Smith but to them.

Charles Barnson, a farmer of Stockport, and James Winchester Scott, a blacksmith from Alma, claimed to be inventor and maker of the first 'good practical plough'. Robert Shapland, of Victoria, said he had invented the principle in 1875. His idea had been for a handle on each share so it could be raised manually. James Martin, of Gawler, said he was the first to put the idea into practical form. Osmund Bull and the Mellor brothers made similar claims.

After a sometimes bitter debate, the South Australian Parliament in 1882

A modern multi-furrow stump-jump disc plough.

awarded Richard Smith £500, a gold medal, and a square mile of land at Ardrossan in recognition of what he had done for South Australia in inventing the stump-jump plough. They conceded that the other claimants may have helped develop the idea, but that Smith was the true initiator. Robert Shapland persisted with his claim, and eventually got £150.

His fortunes restored, Richard Smith joined with Clarence and J. W. Scott and some Adelaide business men in patenting a new version of the stump-jump plough. In 1884 he moved to Western Australia, and set up an implement works at Beverley to make the ploughs.

Clarence stayed on at Ardrossan, gradually improving the plough, and producing models with up to six furrows. He died in Adelaide in 1901, aged 45.

Soon after Clarence's death, some Ardrossan people decided to erect a plaque in his honour, referring to him as the inventor of the stump-jump plough. This brought an angry letter from Richard saying that he and not his brother had invented it. Clarence, in fact, had been nothing more than his apprentice, and had 'expressed strong opinions against the plough, both orally and in writing'. The promoters of the memorial said they did not claim that Clarence had invented the stump-jump plough, but he had certainly made the first one, and done most of the work on development ever since. The marble memorial plaque that subsequently appeared in the Ardrossan Institute was in honour of Clarence Herbert Smith.

When Richard Smith was 74 he went to live in Perth. He died there in 1919, aged 81. He was born in London but came to Australia with his parents in 1838, when he was a year old. Clarence, who was 18 years his junior, was born in South Australia.

Without the stump-jump plough, most of the mallee areas of South Australia, Victoria and New South Wales may never have been cultivated.

Various adaptations of the stump-jump principle have been made since the Smith brothers' day. And when discs replaced mouldboards for ploughing, James Garde of Victoria adapted the stump-jump principle into the Sundercut Stump Jump Disc Cultivating plough. The famous Sunshine Harvester Works, near Melbourne, began producing it about 1906.

The stump-jump plough spread throughout the world, and was particularly useful in the sagebrush and other scrub lands of the western United States. The United States Forest Service used the stump-jump principle in its 'Brushland Plow' of the 1950s.

The Ardrossan firm of Clarence H. Smith Limited carried on making a wide range of agricultural machinery in Ardrossan until 1934, when, unable to collect debts, the company was forced to close down. The factory was demolished, but the powerhouse that once drove it survives in the hands of the National Trust.

7

Louis Brennan's 'ardour for inventions never failed him'

LOUIS BRENNAN was one Australian inventor who did not have to suffer years of blood, sweat and tears only to go unrecognized. When his coastal defence torpedo was tested in Hobson's Bay, near Melbourne, in 1879, it 'excited wonder and approbation'. And the British Government subsequently paid him £110 000 for the rights to it, to prevent other countries from capitalizing on it.

Brennan was born in Ireland in 1852, and at the age of nine went with his parents to Melbourne, where he later worked with a watchmaker and as an engineering apprentice. His interest in things mechanical led him to invent, at the age of 22, a torpedo that could be steered even after it had been launched, and which could be recovered if it missed its target.

Brennan's torpedo was driven by two propellers rotating in opposite directions. Each propeller was driven by an internal drum, this being driven in turn by an unwinding fine steel wire. Each of the two wires was drawn on to a drum on the launching ship or on shore by a 20 horsepower steam engine. Steering the torpedo was simply a matter of changing the speed of winding of one of the wires.

Brennan had his calculations checked by W. C. Kernot, who was then a lecturer at the University of Melbourne and who was subsequently the well-known Professor of Engineering. With the aid of a £700 grant from the Victorian Government, Brennan had a working model made. It was this that 'excited wonder and approbation' when it was tested. The invention was of particular interest to the Victorian Government at the time because of its preoccupation with the naval defence of Melbourne, which it constantly feared would be attacked by the Russians.

Brennan had patented his torpedo on 4 September 1877, before it was tested. When the British Government heard of it they asked Brennan to go to England. Tests of the torpedo by the Royal Engineers at Chatham led to Brennan being awarded £5 000 plus a further £1 000 a year while he continued his experiments. Five years later the War Office accepted the torpedo, and Brennan was granted £110 000, a very substantial sum at the time.

Such a large grant caused a stir in the press, particularly in view of claims that military men had done much to improve the invention at government expense. In spite of moves in the House of Commons to reduce the grant to £30 000, the original figure stood, although £500 of it went to Kernot.

As well as the lump sum payment, Brennan got a good job out of his invention, that of superintendent of the factory the government established at

Louis Brennan.

Gillingham, Kent, to build the torpedo. The War Office was so pleased with the new invention that at first it refused to supply twelve torpedoes to the Victorian Government.

Brennan went on to work on a monorail locomotive with gyroscopic stabilizers, based on an idea he had formulated in Australia. He believed the monorail form of construction would greatly reduce the cost of building railways suitable for mountainous or sparsely populated areas. A model was shown at a meeting of the Royal Society in 1907, and it was publicly demonstrated at Gillingham in 1909.

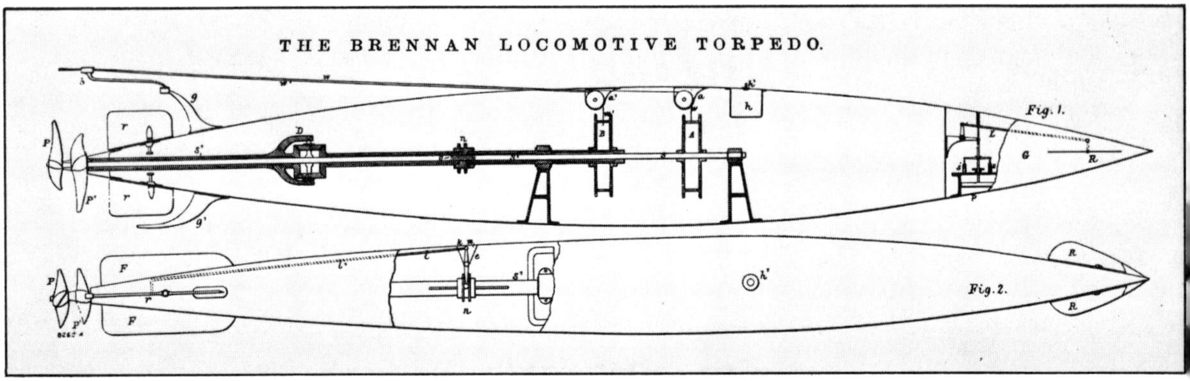

The Brennan Torpedo. The propulsion wires were wound on the two drums (A and B) in the centre of the torpedo. One drum was attached to the inner propeller shaft, the other to the outer (hollow) shaft. When the engine on shore pulled the two wires it caused the drums, and therefore the propeller shafts and the propellers, to rotate.

Although Brennan's monorail train would have travelled at twice the speed of those already in service, the British Government decided not to develop it further, on the ground that people would not ride in a means of transport that depended entirely on one support.

Brennan worked on a series of other inventions, including a biograph that produced a moving picture from a series of drawings or photographs, in the style of a 'what-the-butler-saw' machine. He also worked on a helicopter, but it crashed during tests in 1925.

Brennan was knocked down and killed by a car in Switzerland in 1926. In his obituary, the *Engineer* said, he 'had all the charm of his race and was much beloved of those who knew him well. His ardour for invention never failed him.'

8

James Alston improves the power of windmills

WHEN THE FIRST European settlers came to Australia, they brought the idea of the windmill with them. Early Australian windmills supplied power to grind grain and pump water from wells.

But Australia's greatest demand for windmills came after the discovery of our vast stores of artesian water from 1878 onwards.

Display of Alston's windmills at the Melbourne Royal Agricultural Show, early in the 1900s.

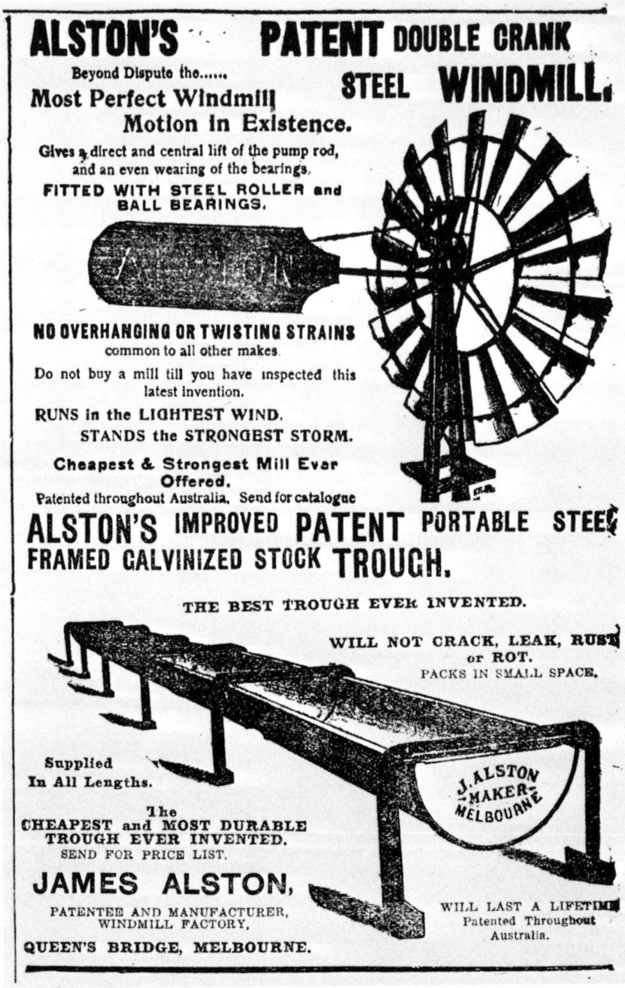

Advertisement for Alston's windmills.

In many artesian bores the pressure of water was so great that it rose to the surface of its own accord. But in cases where it needed help, windmills were ideal. Since many bores were in isolated areas there was no power readily available except the wind's.

Many firms went into the business of making windmills, and naturally many people gave much thought to how they could be improved. In Victoria, James Alston patented an all steel circular windmill with curved sails, which gave much greater power. This is the form still used today by all Australian windmill makers.

To make windmills as maintenance-free as possible, other Australian workers designed models with wind wheels over ten metres in diameter, so that one turn of the wheel would power one stroke of the pump. Smaller windmills can require up to four turns of the wind wheel for one stroke of the pump, and consequently need a system of gears.

9

Hugh McKay made a million from the Sunshine Harvester

As a youth Hugh McKay did his share of winnowing, turning the heavy handle that created a blast of air to separate chaff from grain. Like most people, he detested the job.

In 1882, when he was 17, McKay decided to do something to take the backache out of winnowing. Reaping and threshing had been mechanized. It seemed only logical to him that winnowing could be mechanized too. He soon developed a method so successful that it became the basis of a giant agricultural machinery business that made him a millionaire. When he died in 1926, he left more than £1 400 000.

McKay was fortunate in the way his family listened to his ideas. Many young men of 17 have ideas for changing the accepted way of doing things. Many of them no doubt are impractical dreamers who eventually 'come to their senses'. But how many ideas of genius have been lost to mankind because their elders have laughed at young men who seemed to have ideas beyond their station?

Hugh McKay was fifth of the twelve children of Nathaniel McKay, a poor farmer of Drummartin, Victoria. Born in 1865, he grew up in a slab hut on the farm.

When he told his father of his plans to build a combined stripper-harvester and winnower, he got a good hearing and every encouragement. With the help of his brother John, the young inventor set about building a prototype. They built a smithy of logs and bark to house the family forge.

Too poor to buy materials, the brothers used whatever they could find from disused machinery, timber and junk metal. They took the land wheels and axles from a vintage reaper to support the frame of the harvester. They trimmed down the comb from a stripper to about half its length to suit the dimensions of the prototype. They took sprocket wheels and chain drives from an old binder, and borrowed riddles from a winnowing machine.

Hugh's brother Nathaniel, the only trained tradesman in the group, showed the boys how to square and set the harvester frame and how to adjust the bearings. John did the carpentry, and another brother, George, the iron work. Their father fashioned and tempered tools needed for the various components. They made pulleys from a red gum log they had to split with wedges, and even gunpowder. Several bearings also came from the same material.

The construction job proved formidable indeed. To the McKay family's great disappointment the 1883 crop matured earlier than usual and they found

H. V. (Hugh) McKay.

it impossible to complete their machine in time for it. But they saved about a hectare of standing grain for trials with the harvester late in the summer. By the end of February 1884 the world's first stripper-harvester-winnower was ready for its critical test.

The comb separated the grain stalks, then a five-bladed beater stripped the heads from the stalks. It swept them against a concave surface which threshed

The farm smithy at Drummartin, where Hugh McKay in 1884 built the first Sunshine harvester.

the grain from the heads by the abrasive action of the beater blades as they ground the heads against the concave.

The resulting mixture of grain, chaff and straw was elevated by a belt-driven slatted conveyor and dropped on to the riddles for separation. Air from a rotary fan blew unthreshed heads over the front of the riddles and down into an elevator to be delivered back to the drum and rethreshed. Straw was blown out under the machine.

Clean grain passed through the top riddle to a grain elevator chute and travelled by a metal cup-type conveyor to a hopper under which a bag was suspended. Small foreign seeds, pinched grain, chaff and dirt fell through the lower sieve into a box with a hinged flap bottom for easy emptying.

When the machine was tested, Hugh McKay found that its main troubles were in the unique winnowing mechanism.

One major headache to the young inventor was the need to provide movement to the riddles, to simulate the action of hand operated sieves. Without this movement, the mass of straw and grain just piled up on top of the riddles. McKay experimented until he solved this by including a crank drive that moved the riddle tray sideways as well as back and forth, to provide lateral shake.

A harder problem was how to ensure that the riddles stayed horizontal regardless of height adjustments to the comb front. McKay solved this in time by fitting an independent hand lever to keep the riddles level.

McKay senior delved into his meagre cash resources and allocated twenty pounds for the entire project. The only known payments were 10 shillings to an Elmore blacksmith to fit a band to the forecarriage and 30 shillings for nuts and bolts. The family bought a few tools, but they made most of them on their farm: an improvised drilling machine, bits of various sizes for the drill, a lathe, chisels, and punches.

They must all have been very excited when they finally took the harvester across to the 'well paddock' for its first field trial on that February morning in 1884.

When they started the machine, two of the belts came off the wooden pulleys, a problem they soon overcame by fitting flanges. At the second attempt there was no hitch, and the harvester delivered clean grain into the bag on the forecarriage.

The McKay family raised enough money to patent the winnowing mechanism. Hugh McKay then began proposing to stripper manufacturers that they should make his machine on a royalty basis. One by one they rejected his overtures, either from complacency, a desire to safeguard their existing products, or sheer conservatism.

The original Sunshine harvester.

Sunshine harvesters awaiting despatch from the factory.

Eventually however McKay made a deal with North Melbourne plough makers, McCalman & Garde, to build five machines, and he went into the business of selling them himself, in time for the 1885 harvest.

A small public company, the McKay Harvester Co., was formed in Ballarat in 1886 to handle manufacturing contracts and sales. The company had no plant of its own, but subcontracted to various firms: Henderson & Co. of North Melbourne; Pickle of Bendigo; Quale & Williams, Munro & Sons and Harrison of Ballarat. But in the depression of 1892 the McKay Harvester Co. was one of many enterprises forced into liquidation.

With the help of a few friends, McKay bought back his own assets and started the business again. Operating in a small shed at Yuille Street, Ballarat, Hugh McKay played the multiple roles of manager, salesman, factory hand and designer.

By 1894 the Yuille Street plant had been extended to 185 square metres. It was equipped with two lathes, a drilling machine, hand punch, grindstone, forge and bellows. A horseworks pulled by two grey ponies provided the power.

This was the last year during which strippers were converted, because Hugh McKay had devised an improved harvester.

The most significant additions to the new machine were a peg drum and screen. Material from the beater was elevated to the peg drum for extra threshing before it was dropped to the riddles for cleaning.

A slotted cylindrical screen on top of the machine diverted foreign weed seeds to a second box to ensure that only clean grain went into the bag. The first twelve of the new machines were built under a contract agreement in 1895 by Jack and George Butt and William Hopkins.

Demand for the harvesters was quite beyond the capacity of the small plant and McKay farmed out work to local engineering firms. Even the introduction of steam power was a mixed blessing, since the factory lacked space for the boilers. McKay overcame the problem by having the boilers across the road and running the steam through underground galvanized pipes. He bought up adjoining cottages for the blacksmiths, cutting-up shop, woodmill, offices, storage and testing sections. Eventually the painting and despatch sections were in an old brewery building. McKay's Ballarat plant was bursting at the seams, a sure sign of success.

McKay chose the name Sunshine for his harvester when Dr Talmadge, a visiting evangelist, gave an address entitled 'Sunshine' in Ballarat.

Eventually the McKay business outgrew Ballarat, particularly when the Victorian Railways refused it any freight concessions. Once McKay developed a substantial export trade, he started to look for a site closer to Melbourne. He found it in 1904 when the Braybrook Engineering Works at Braybrook Junction met financial difficulties and was offered for sale by tender. It was right at the junction of the Ballarat and Bendigo railway lines.

McKay's bid was accepted and in 1906 he started to transfer his business to Braybrook Junction, which was renamed Sunshine in 1907.

McKay's business grew steadily and made a wide range of agricultural machinery in addition to the Sunshine Harvester. Remembering perhaps his own early frustrations, McKay was always willing to help other inventors. In particular he adopted the ideas of Headlie Taylor, who invented an improved type of harvester in 1915. Taylor joined McKay's firm and stayed with it for the rest of his life.

During World War I, McKay converted his factory to the manufacture of transport and ambulance waggons, watercarts, portable kitchens, trenching tools and munitions.

The plant he founded at Sunshine is now operated by Massey-Ferguson (Australia) Ltd. It still makes versions of the Sunshine Harvester. The smithy in which Hugh McKay made his first harvester is preserved there.

10

Putting the screws on wool

WOOL REALLY WAS a natural for Australia. Compared with a product such as wheat, wool represented great value in small bulk. It could stand the economics of bullock waggon transport from outback stations to the coast, and the long sea voyage from there to the woollen mills of England.

It could also be compressed, to get even greater value into the same bulk. So some kind of wool press became an essential piece of shearing shed equipment.

At first the shed hands laboriously tramped the wool down, starting with a

Baling wool the old way, by stamping it down.

John Jordan with the Autoflow automatic wool press, designed and made by his firm in Western Australia.

long bale and finishing with a short one. An arrangement of wire ropes and a steel plate called a monkey replaced this, but getting the wool into the bale still called for a heavy expenditure of sweat.

The big breakthrough came when somebody fitted a rack and pinion to apply pressure to the monkey by turning a handle. An early version of this type of wool press was patented in Queensland in 1877.

One of the best known names in improving the wool press was that of Christian Koertz, a Danish mechanic who settled in Sydney in 1887 and set up in partnership with Frederick Mason, who held patent rights to a press. Using

Koertz's natural flair for invention, they turned out a steady flow of improved wool and hay presses.

The partnership was dissolved in 1913 and Koertz continued in business on his own account, making wool presses for both large and small sheep men. His 'Squatter's Press', 'Little Wonder', 'Improved Selectors', and 'Homestead Lessee's' presses met with steady demand. Using a Koertz press, one man could exert 30 tonnes of pressure on the baled wool.

Most wool presses have two boxes joined by hinges. The bale is fitted into the box that will stay on the floor, both are filled with wool, and one is lifted or swung on top of the other. Then the monkey plate forces all the wool from the top box into the bottom box and, hence, into the bale which is then sewn up.

Christian Koerstz invented many improvements to wool presses.

Work on improving the wool press has continued steadily, right up to the present day. The Autoflow wool press, which won the new implement award at the Australian showplace for new rural machinery, the Orange Field Days, in 1977, is built for lightness, so one man can move it, an important consideration in these days of high cost labour. The Autoflow has three opening doors and a set of rollers, so full bales can be removed easily. It also has a built-in scale so bales can be filled to the maximum weight without incurring penalties for being overweight.

Modern wool presses are driven by electricity, or where that is not available, by petrol engines, which makes them a far cry from those in which the wool had to be forced into the bale by nothing more than brute strength with perhaps a little help from a lever.

11

Experts said George Chaffey's pumping engines would never work

---

By the late 1880s Mildura irrigation pioneer George Chaffey had learned to live with critics. He'd been branded a lunatic when he bought the rabbit-infested, drought-ridden Mildura station on the Murray river and announced his intention of turning it into an oasis of orange trees and vines. And he'd been branded a 'Yankee adventurer' when he'd propositioned the Victorian Government to grant him land so he could start an irrigation colony there.

But he'd managed to overcome all obstacles placed before him, so when world experts declared that his design for triple expansion direct steam pumping engines simply would not work, Chaffey decided to go ahead and build them anyway. Not only did they work in the way Chaffey intended, but they kept on working for the best part of 70 years.

Born in Canada in 1848, Chaffey spent his early years in his father's ship-building business in Ontario. Here he designed steamships that became famous for their combination of speed, light draught and carrying capacity. In his early thirties he moved to California, where he became President of the Los Angeles Electric Company, which made Los Angeles the first city in the United States to be lighted exclusively by electricity. He installed the first long-distance telephone in California. But it was in establishing irrigation colonies that Chaffey really made his reputation, and attracted the attention of Alfred Deakin, a future Prime Minister of Australia who in 1885 was the Victorian Government Minister in charge of water resources. Deakin persuaded George Chaffey and his brother William to come to Victoria. The present irrigation areas around Mildura and Renmark are a direct result of the Chaffeys' work.

By the late 1880s, work at Mildura had advanced to the point where engines would be needed to drive the pumps that would pour water into the irrigation channels. George Chaffey designed triple expansion marine steam engines, and proposed that they be connected directly to centrifugal pumps without benefit of belts, ropes or gearing of any kind.

Triple expansion steam engines were then working in the largest P & O liners, but they had never been tried before in irrigation, and certainly not as

Chaffey proposed to use them. Engineering manuals in Britain and America ran special articles about the likely efficiency of Chaffey's engines. The British builders, Tangyes, agreed to make the engines to Chaffey's design, but refused to accept responsibility for them. They played down their role by labelling the engines:

George Chaffey.

'The Chaffey Improved Pumping Plant, made by Tangyes for the Mildura Irrigation Colony'.

Once they were installed, the two engines worked perfectly and pumped water almost endlessly for Mildura and district until electric motors replaced them 70 years later. At the time they were built the pumping stations at Psyche Bend and Billabong were the largest in the world. Sir Arthur Tangye was so interested in Chaffey's engines that he travelled half-way round the world to see them in action.

Chaffey's triple expansion direct drive engine now stands on the banks of the Murray at Mildura as a tourist attraction.

George Chaffey's business interests in Australia suffered severely in the depression of the early 1890s, unleashing the knockers once more. Fed up with Australia, George Chaffey went back to North America and carried on his work in irrigation there, with outstandingly successful results. He died in 1932.

His brother, William Benjamin, usually known as W. B., spent the rest of his life in Australia, living at Mildura in 'Rio Vista', the mansion he built and which today houses a museum of the history of irrigation in the Sunraysia area.

12

Lawrence Hargrave walked on water and unlocked the secret of powered flight

'IF ANY MAN deserves to fly, that man is Lawrence Hargrave of Sydney, New South Wales', wrote American pioneer aviator Octave Chanute at the close of the nineteenth century.

Hargrave had by then unlocked the secret of powered flight by building a series of kites which hauled him off the ground in a strong wind. With a little more encouragement from his associates and a little more persistence on his own part, Hargrave might have become the first man to make a powered flight. But most of his fellows in the Royal Society scoffed at his ideas, causing him to retreat into himself and become too preoccupied with ornithopters, flying machines with flapping wings.

Born in London in 1850, Hargrave was only six when his parents dumped him in a boarding school in England and took off for Australia. He did not see them again until he too landed in Sydney ten years later. He disappointed his father, a prominent judge, by showing no talent or inclination for following him into the law. Everybody who knew Hargrave could tell that his aptitudes were better suited to engineering, but his father persisted. Only the boy's failure to matriculate put an end to the idea.

The young Hargrave became an apprentice in the workshops of the Australian Steam Navigation Company, learned how to design on the drawing board, and became an outstanding craftsman in wood and metals, particularly tin-plate.

Hargrave's inventive urge soon began to show. In 1870, when he was 20, he devised a means of walking on water, and was often to be seen plodding across Rushcutters Bay, each foot in a boat-shaped boot long enough to make him float. Each boot had a system of hinged flaps beneath it. As Hargrave moved each foot forward these lay flat against the bottom of the boot, but as soon as it stopped they became vertical and reduced backwards movement. Hargrave could take short cuts across the bay, but other people never seemed to get the knack of walking on water, and his invention came to nothing.

In his twenties Hargrave made several journeys to New Guinea. He was

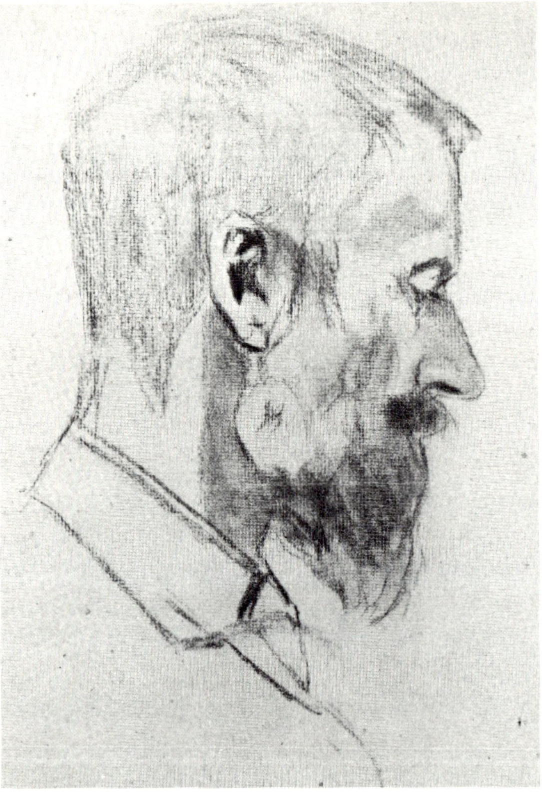

Lawrence Hargrave about 1910.

aboard the ill-fated brig *Maria*, bound there with settlers, when it foundered on the Barrier Reef, with considerable loss of life. Later he explored the Fly River as one of a party led by Italian naturalist D'Albertis, and he was sent by Parbury Lamb and Company to sort out the problems of a drunken and thieving superintendent at their pearling station.

Back in Sydney, Hargrave began work as an observer at the Sydney Observatory in 1878 and stayed there until 1883, by which time gifts of land from his father had made him a man of modestly independent means. He gave up his paid employment so he could devote himself full time to the experiments on which he had already been spending most of his leisure.

Hargrave had become preoccupied with thoughts of flying. In 1870 he had been one of the large crowd that watched the aeronaut Gale take off in a balloon from the Sydney Domain. Soon after this he sketched an improved balloon, which he later called his 'Flying Machine No. 1'. It had a long cigar shaped gasbag with a rudder at the stern and was driven by an airscrew turned by hand through a system of geared pulleys.

Hargrave's particular interest was his 'trochoided wave theory', a trochoid being the geometrical shape described by a point on the radius of a rolling circle. By studying birds and animals, particularly snakes, fish and worms, Hargrave formalized his theory, and even built a working model of an earthworm to demonstrate it. He showed that the forward movement of many animals depended on a wave-like motion, and he hoped to use a similar motion in his flying machines.

Like most people who tried to design early flying machines, Hargrave at first put his hopes in flapping wings, particularly flapping wings that reproduced the trochoidal motion of a bird's wings. His first model ornithopter to achieve free flight did so on 31 December 1884. It weighed one ounce and was driven by two flappers powered by two stretched rubber bands. It was not a new invention, for other experimenters had made similar models in Europe as early as 1858.

Hargrave continued to develop his models. One, its flappers powered by 16 rubber bands, flew 33·5 metres. He even made one ornithopter with a hand-driven mechanism that flapped its wings in a figure of eight motion. When he wound the handle the ornithopter shuddered and shook from side to side but would not leave the ground.

Convinced by the failure of his hand-driven mechanism that successful flying machines would need extra power, Hargrave began to take an interest in engines. In 1888 he designed the first of 36 he was to build, a single cylinder compressed air motor driving two flappers.

His third design was for a radial rotary engine, the forerunner of the type eventually used in so many aeroplanes, and eventually superseded only by turbines and jets. As usual, he declined to patent it. Years later, the Seguin

Hargrave's last india-rubber propeller-driven flying machine flew on 31 January 1889.

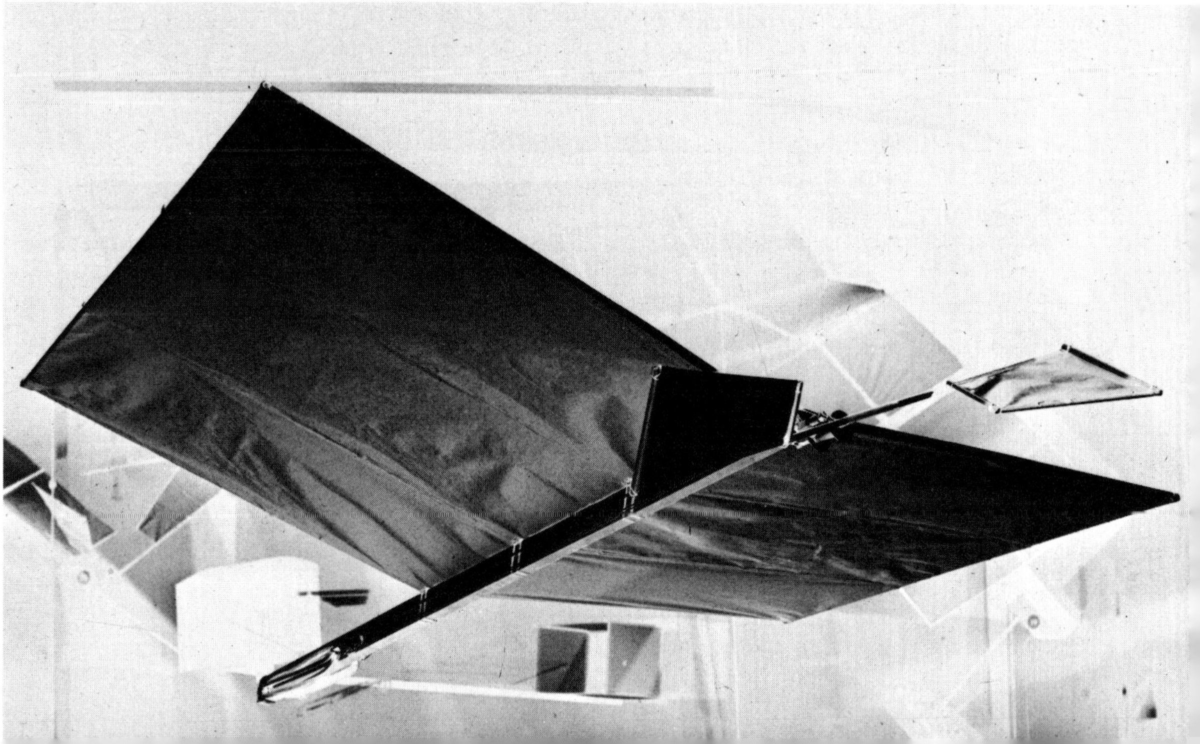

brothers in France built engines based on a very similar principle, the famous Gnome, Clerget and Le Phone rotary engines that powered so many planes during World War I.

For a time Hargrave switched to fixed wing models powered by propellers of various kinds. On 31 January 1889 his last india-rubber-powered propeller-driven model flew 33·5 metres before it crashed. He then switched to engine-powered models and his Number 13 flew 40 metres in eight seconds on 2 January 1891. It was a propeller-driven monoplane powered by a three-cylinder rotary engine.

If he had persisted with this approach he may have been more successful in the long run, for we know now that the major defect was that his propellers were set at the wrong pitch. Continued experiments may have overcome this problem, but Hargrave turned away from the idea of propeller-driven flight and went back to flapping wings.

When Hargrave's brother Ralph died in 1888 he left him 'Hillcrest', a house

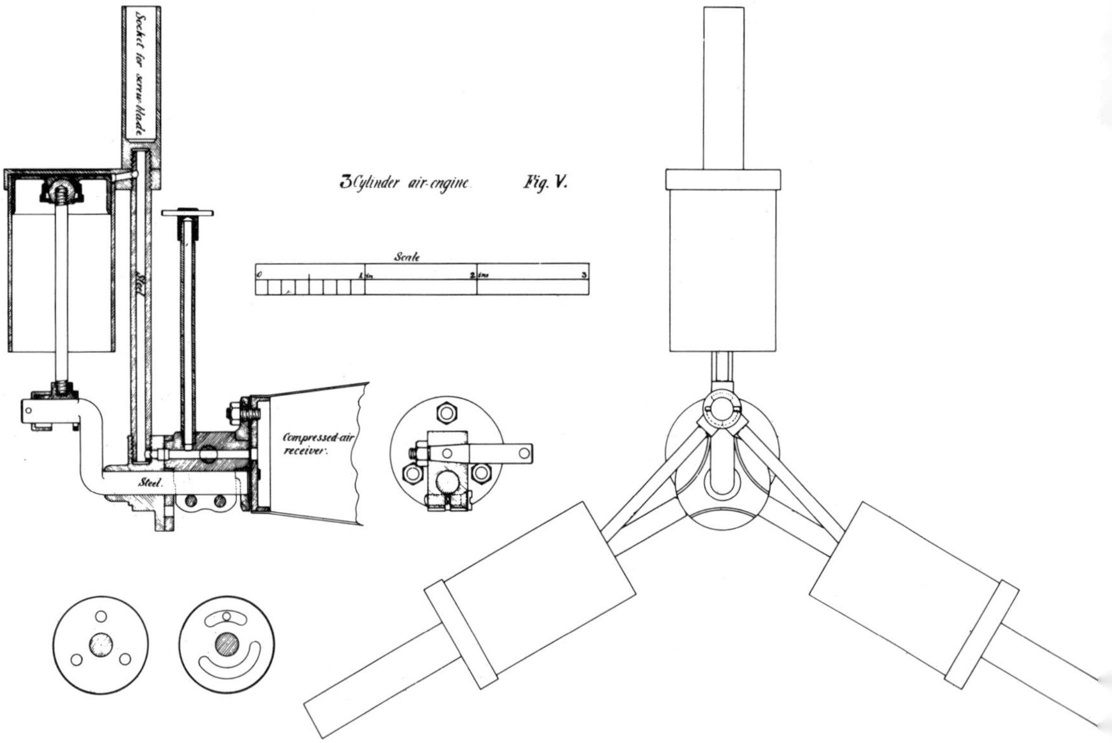

Hargrave's drawing of his three cylinder radial rotary engine.

Hargrave with the kites that lifted him 5 metres off the ground at Stanwell Park on 12 November 1894.

at Stanwell Park, on the coast south of Sydney. Hargrave moved there with his family in 1893 and carried on his work.

He had by then begun a new series of experiments, aimed at testing the lifting power of curved surfaces.

He found that a segment of a paper sphere placed with the convex side up provided lift, but when reversed it pressed down. He also built a windmill with two sets of sails, one with curved sails and one with flat. With the sails set in opposite direction the curved ones easily overcame the flat. But Hargrave at first failed to realize that the lift was provided by the airstream pulling the convex side rather than by pushing the concave one, and that the wing was not supported by the air beneath it.

To further his experiments on the lifting power of surfaces Hargrave began to use kites, the first of which he flew in Sydney on 21 January 1893. Most of his later experiments were at Stanwell Park, where on 12 November 1894 he linked four kites together to lift himself 4·8 metres above the ground in a 33·8 kmh wind.

Hargrave moved on from kites to a series of gliders, which he called soaring machines. He effectively demonstrated the extra lifting power of the deep parabolic

Early European aeroplanes. Notice the resemblance to Hargrave's box kites.

wing, and probably influenced the all-important wing design of the successful Wright brothers gliders. The Wrights also used the thickened leading edge developed by Hargrave.

Meanwhile, Hargrave continued his work on developing an efficient engine that could give adequate forward motion to his proposed flying machine. For a time he concentrated on steam engines fired by methylated spirits, kerosene and naphtha. In six years of experimenting with steam engines he built five cylinder-based engines and five others of varying types, all of them failures. He designed nine more which he did not build.

As well, he designed a turbo-jet engine and actually built a pure jet engine, but could not get it to work.

In 1899 Hargrave went to England and spoke before the Aeronautical Society of Great Britain, but could not get more than token support for the development of his ideas. He continued his work even after the Wright brothers flew successfully for the first time in 1903, but in 1909, overcome by lack of money or other support, he gave up.

Hargrave had a rare combination of talents. His analytical and inventive brain devised a constant flow of new ideas, his talent for draughtsmanship enabled him to draw them in detail and his skill as a tradesman enabled him to give his drawings expression in finely constructed models.

His major fault was his lack of talent for communicating his ideas and enthusing people whose support he could well have used. His explanations to the Royal Society were not always as interesting or exciting as the subject warranted. His highly technical explanations left his audience wondering what he was really trying to tell them. They usually listened patiently enough, but were inclined to shake their heads behind his back.

Hargrave's usual reaction to criticism or indifference was to withdraw within himself, and get on with his experiments. He was working in a vacuum. Unable to swap ideas with his fellow researchers overseas, he may well have benefited from a wider circle of people with whom he could discuss his work in Australia. Who is to say that a friendly critic to whom he listened may not have encouraged him to go on with his propeller experiments?

Even after he gave up his experiments in flight, Hargrave kept up his interest in other fields of science and exploration. He developed an elaborate theory to prove that Spanish explorers had visited the east coast of Australia long before the British. He drew up a plan to turn Sydney into an even greater port than it already was.

Hargrave's final tragedy came in 1915, when his beloved only son and fellow experimenter, Geoffrey, was killed at Gallipoli. Two months later he too was dead. Some people said the same bullet had killed them both.

13

Coolgardie safes cooled goldfields' tucker

When thousands of people joined the gold rush to Coolgardie, Western Australia, in the 1890s, they needed some way to keep food cool. They were a long way from ice or refrigeration.

Somebody remembered that evaporation absorbs heat and that in Coolgardie's hot, dry climate water evaporates readily. So the Coolgardie safe became

A commercially made Coolgardie safe of 1917.

commonplace. Just who thought of it first is anybody's guess. It may even have been around before the Coolgardie rush, but it certainly was standard domestic equipment on the Western Australian goldfields.

The usual Coolgardie safe was an open timber frame with a tray of water on top. Strips of hessian or other fabric with their ends in the water hung down the

A commercially made canvas butter cooler, which worked on the same principle as the canvas water bag.

side. Water percolated down them, evaporating and absorbing heat as it went. The safe worked best if it was in a breeze, so it usually stood on the verandah, with its feet in tins of water to keep ants away.

The Coolgardie safe was a fairly logical development of the canvas water-bag, which Australians had been using for about 40 years before they even heard of Coolgardie.

Outback explorer Sir Thomas Livingstone Mitchell, is credited with inventing this. Observing that Aborigines carried their water in kangaroo skins, he realised that they had a far better method than the wooden kegs he was using.

He had some canvas bags made, and greased them with tallow to make them watertight. But tallow tainted the water, so he tried plain canvas. Once the canvas was wet enough it held water satisfactorily. There was a bonus, too, for water seeping through the canvas to the outside evaporated and cooled the contents. Like the Coolgardie safe, the waterbag worked best if it were left in a breeze.

14

From 'Megaethon' to 'Big Lizzie': Attempts to go where the roads do not

Moving a heavy vehicle across country can have its problems. Unless the ground is very solid, the vehicle is likely to dig itself in and resist all attempts to move it further. Mud, sand and dust can all cause this.

The obvious solution is to increase the bearing surface in some way, so the wheels are less likely to dig their way into the ground under the vehicle's weight. If the vehicle can lay its own bearing surface in front of itself, then pick it up again after it has passed over it, so much the better.

One early attempt at this approach occurred in 1850 when a Hunter Valley farmer named Cleve had a steam engine built in Sydney. He decided to take it home under its own power on the iron shoes it laid down in succession in front of its wheels. Cleve called his machine the 'Megaethon', but the Aborigines who saw it called it the 'buggy buggy'.

Eventually Cleve had to get some bullock teams to come and rescue his engine. Its great weight made its progress very difficult.

The idea cropped up again when steam-driven traction engines became common in Australia, about the turn of the century. Used to haul waggons across country, for ploughing and for stationary power at all sorts of locations, these machines faced the need to travel across country in all sorts of conditions. In 1906, Frank Bottrill patented 'an improved road wheel for vehicles and travelling machines, especially useful for traction engines'. He called it the Pedrail or Dreadnought wheel.

As Bottrill's wheel rotated it placed a series of bearers one at a time upon the ground. Each bearer formed a substantial flat bed while it was on the ground and prevented the wheel from sinking. The engine was therefore able to exert its full tractive effort. For use in very loose ground the bearers could be fitted with studs.

Some versions of Bottrill's wheel had two sets of bearers side by side. The bearers were attached to the wheel by a system of U-bolts and wire ropes that allowed them to move in relation to the wheel, but kept them rotating with it.

In 1907 Bottrill used a 30-horsepower International tractor fitted with his

Big Lizzie in her last resting place at Red Cliffs, Victoria.

Pedrail wheels to plough 202 hectares of mallee land for the Victorian Department of Agriculture and 38 hectares of virgin country at Howard's Plain, Victoria. Then he used a 70-horsepower McLaren traction engine with Pedrail wheels on a large-scale land-clearing operation at Tintinara for the South Australian Government. He pulled three large rollers covering a span of 18 metres and cleared an average of 12 hectares a day, and sometimes as many as 20 hectares. Later the South Australian Government bought rights from Bottrill to fit two of its steam tractors with his wheels. The Queensland Government also arranged to fit them to some of its equipment.

In World War I, the Australian Light Horse in Egypt had the Pedrail system fitted to its field guns so it could haul them across the desert. The heavy sand made transport of the guns on their ordinary wheels quite impracticable, and for a time General Chauvel was forced to conduct a mounted campaign against the Turks without artillery support. Once the Pedrail was adapted for desert conditions and fitted to the guns, the problem was overcome. Guns fitted with Pedrails were first used in the attack on Salmana in May 1916.

Bottrill's best known application of his Pedrail was to Big Lizzie, a giant traction engine now on display at Red Cliffs in Victoria's Sunraysia district. The massive machine, powered by a 60-horsepower Blackstone crude oil engine, was built in Melbourne in 1914. Its range of gears gave it four forward speeds from 0·8 to 3·2 km/h, and two reverse speeds, 0·4 and 0·8 km/h.

Until Bottrill style wheels were fitted to artillery in the Middle East during World War I, the Australian Light Horse had to attack fortified Turkish positions without artillery support.

Big Lizzie set out from Melbourne in 1915 and went via Echuca, Kerang, Swan Hill, Ouyen and Mildura. Everywhere she went, her owners had to get permission from the various shires. From Ouyen she travelled along a bush track made by bullock waggons beside the railway. This was the only road. Where the turns in the track were too sharp for her 61 metre turning circle, Big Lizzie just made her own track through the mallee scrub. Her big wheels carried her across even the sandiest country. She had a six weeks stop-over in Kerang while all her wheels were taken off and altered.

Big Lizzie arrived in Mildura in October 1917. She was unable to cross the Murray, which was in flood. No bridge or punt could carry her, so she went to work in the district, carrying wheat, one 1919 load running to 900 bags.

Big Lizzie herself was 10·2 m long, 3·4 m wide and 5·5 m high. She had two flat top trailers, each fitted with Bottrill wheels. Each trailer was 10 m long, 3·4 m wide and 2 m high.

When the Victorian Government decided to make farms for soldier settlers in the Sunraysia area during and after World War I, Big Lizzie helped clear land at South Merbein, West Merbein, Birdwoodton and Red Cliffs, a task that lasted until 1924. She cleared land in other parts of Victoria until 1929, when she was abandoned.

She was finally brought back and given her place of honour at Red Cliffs.

15

George Julius turned failure to success with his automatic totalizator

When western australian politicians rejected his mechanical voting machine, George Julius was only temporarily downcast.

He soon adapted his mechanism to serve a great need, something that would take over the mass of calculations involved wherever people gathered to bet by the totalizator system. What the politicians rejected the racing fraternity soon embraced, and the Julius style of automatic totalizator became a standard fixture on racecourses all over Australia and around the world.

Born in England in 1873, Julius came to Australia as a child with his clergyman father, from whom he obviously inherited his love of things mechanical. The senior Julius was well-known wherever he went as a fixer of clocks, and he was famed for the magnificence of his home workshop.

George Julius graduated in engineering and became an assistant engineer with the Western Australian Government Railways in the 1890s. When the losers in an election there claimed the voting had been irregular, he set to work and invented a fool-proof voting machine. It worked well enough until the government changed the voting system. Julius adapted the machine, but he could not persuade the politicians to adopt it. It just stood silent and idle in his workshop.

Although he had no particular interest in horses, Julius began to feel that he may be able to adapt his voting mechanism to replace the small army of clerks needed for the Pari-Mutuel betting system invented in France in the 1860s. The system was popular in Australia, where it was known as 'the tote'.

There already were tote machines, but they were far from automatic. As the ticket sellers took bets they called them aloud to a man at a board behind them. On this board was a series of numbers, each with a lever beside it. A man pulled the appropriate lever as the ticket seller called each bet.

Julius wanted to make a fully automatic machine that would record the bets and calculate the dividends too. He was still working on his machine when in 1907 he moved to Sydney to establish his own business as a consulting engineer. One of his first acts there was to set up his workshop, where he continued to work on his tote, an electrically operated maze of wheels, cogs, piano wires and lead weights.

The first racecourse to accept the Julius automatic totalizator was Ellerslie, in Auckland, New Zealand, where it went into action one Saturday in 1913. Preprinted and serially numbered tickets were issued by hand to acknowledge bets, but from there on the machine did everything else automatically.

Sir George Julius, inventor of the automatic totalizator and later first chairman of the CSIRO.

One early visitor to see the new tote in action was the inventor's father, Churchill Julius, by now His Grace the Anglican Archbishop and Primate of New Zealand. He stayed so long looking at the tote that the races crowd had gathered by the time he left. There was already a rumour that the Archbishop had really designed the tote, and that because of his calling he had chosen to let his son have the credit. His presence on the very first day the tote was in action only added to the rumour. 'That new-fangled tote machine has broken down and they have had to get the Archbishop in to fix it,' people knowingly said to each other.

Back in Sydney Julius formed Automatic Totalizators Ltd and put his tote on the market. Gloucester Park racecourse in Perth installed the first tote in Australia, in 1916. The main Sydney racecourses installed them in 1917-18, but Melbourne held out until 1931. Australian designed and made totes spread to just about every country in the world where horses race.

The usual tote prints the ticket, sends the betting data to the various pools, makes deductions for taxes and operating costs, and instantaneously adjusts the odds shown on the public indicators. Julius continued to improve the machine and in 1932 invented an automatic odds-computing device, first used in Melbourne, which eliminated any time-lag in showing on the public indicators all changes in the order of favouritism of runners.

The day the first automatic totalizator operated publicly. Ellerslie Racecourse, Auckland, New Zealand, 22 March 1913. Betting data appeared automatically on the face of the central building.

Mechanism of an early automatic tote.

Julius had far more widespread scientific and engineering interests than the tote. When the Bruce-Page government established the Council for Scientific and Industrial Research (later the CSIRO) in 1926 it made Julius the first chairman, a position he held for 20 years. Julius was knighted in 1929 and died in 1946.

Interest in things mechanical lived on in the family. Sir George Julius' son, Awdry, is a Sydney consulting engineer. He has maintained the family connection with Automatic Totalizators Limited, of which he is chairman.

16

A German U-boat prodded the British Admiralty into accepting Michell's thrust bearing

EARLY IN WORLD WAR I the Royal Navy captured a German U-boat intact and found it equipped with an impressive new thrust bearing in which a film of oil completely eliminated metal to metal contact.

Because of the greatly reduced friction to be overcome within the bearing, the U-boat used considerably less fuel than it would have done with a conventional bearing. And the new type of bearing took only a fraction of the space of a conventional one.

Fuel economy and space saving were both very important in submarines. Both were also important in surface ships, and the Admiralty soon began fitting its ships with the new type of thrust bearing.

British ship designers learned from the German Navy what an Australian inventor, Anthony George Michell, had been trying to tell them for the best part of 10 years. Michell had patented his invention in 1905 but could not persuade manufacturers to use it.

Born in Victoria in 1880, Michell was the son of English immigrant parents. He lived in the bush around Maldon for most of his childhood, until he went to Melbourne and Cambridge to study for his chosen career in engineering.

As a consulting engineer in Melbourne he worked on the designs of large pumps for Murray River irrigation schemes. In this work he saw the limitations of conventional thrust bearings, in which there was metal to metal contact. He started to think there must be a better way to make bearings, and his interest turned in particular to the sort of bearings used on the main propeller shafts of ships.

At the turn of the century, a single propeller shaft in a large ship might have up to fourteen collar bearings. These were necessary to take the thrust of the rotating propeller, which could push the ship along only by pushing against something attached to the ship.

Michell studied Reynold's theory of lubrication, published by the Royal Society in 1886. He developed Reynold's mathematical treatment of the problem of viscous flow in a finite space, and in 1905 he published in a German journal on mathematics and physics his famous paper on the lubrication of plane surfaces.

Michell showed how the problems of the flow of lubricating oils could be treated on a scientific basis. He used this theory as the basis for his design for a new thrust bearing in which the bearing surface is a series of sector-shaped pads, each of which is free to tilt so a wedge-shaped film of oil can form. In a bearing

Anthony George Michell, 'a modest man of genius'.

made to this design, the friction is no more than that due to the viscous flow of the oil and is almost negligible. Michell's bearing was shorter than previous ones, and this saved space.

Until the Admiralty adopted the idea in 1915, Michell could get no recognition for his idea. Few engineers knew much about the theory of lubrication and were sceptical about Michell's claim that his bearings could handle the loads involved. German submarine designers, however, had seen the merits of the idea, but they were not about to tell the world of Michell's marvellous breakthrough.

When Michell applied for an extension of his patent in 1919, he proved that the invention caused a very great reduction in friction and therefore increased the power transmitted by the shaft to the propeller. Its use in ships of war had led to a great saving in coal and oil, in the cost of construction and in the space occupied by the bearing.

In 1920 Michell formed a company and set up a workshop in the Melbourne suburb of Fitzroy, where he made prototype crankless engines. The company's ultimate aim was to license manufacturers abroad to make Michell crankless engines.

Michell's great hope was to have his engine used in cars. In 1923, he had an eight-cylinder crankless prototype built and taken overseas for demonstration. A second engine was installed in Melbourne in a Buick chassis.

The overseas mission was led by a young, personable and persuasive engineer, Richard Casey, a director of the company and a future Governor-General. Ford and General Motors both agreed that Michell had invented a very good engine, with a performance about 10 per cent better than the standard car engines of the day. But this was not enough to warrant the cost of retooling and re-equipping.

Michell's Melbourne company closed down in 1928, but there still seemed to be some hope for his engine in the field of aviation. The Smithsonian Institution in Washington has two Michell type aero engines that were built for the United States Navy, one of them as late as 1942.

Michell thrust bearing: pads on left and second part of thrust on right.

Eight-cylinder Michell automobile engine in Buick chassis.

Eight-cylinder Michell petrol engine. On the bottom of the picture is the reciprocating double-ended piston which drives the shaft through a tilted or inclined disc.

Eventually, Michell's plans for a crankless engine were never a commercial success. S. E. A. Walker, an engineer who as a young man worked with Michell in his Melbourne workshop, has diagnosed the failure as 'too few people trying to do too much with too little capital'.

'Michell, despite his world renown,' Mr Walker says, 'had no great corporation to back, with adequate resources, his great invention. With major accent on the design of many machines, the funds for promotion were limited.'

The coming depression made car manufacturers wary in the late 1920s, when Michell was trying to get his engine accepted. Development of rotary and jet engines in aviation and the phasing out of the reciprocating engine for large power aero engines called a halt to the development of crankless aero engines.

Walker has described Michell as a 'modest man of genius'. He was 'a man of such rare intellectual quality, logical, patient, completely without condescension, always helpful and understanding.'

Perhaps all that is why he never made a commercial success of his crankless engine. How many successful businessmen do you know who would fit that description?

17

Potter and Delprat put froth and bubble to work on Broken Hill's elusive zinc

THE NEW SOUTH WALES Mines Department reported in 1901, 'The future of Broken Hill is dependent on two things, the price of lead and the discovery of some successful method of treating zinc. The zinc product is dumped on the mines, where it remains a vast asset, locked up. The zinc problem at Broken Hill is therefore the question of the hour.'

By 1903, the Broken Hill tailings dumps contained metal worth about £30 million, if only it could be recovered.

When boundary rider Charles Rasp first found the mineral outcrop at Broken Hill in 1883, he had thought at first that it was an ore of tin. In fact it was one of the largest deposits of silver-lead-zinc in the world.

Broken Hill's mining companies soon worked out how to recover up to 70 per cent of the lead and up to 50 per cent of the silver from the ore, but they still did not know how to extract the zinc. By the turn of the century, they had stockpiled 6 000 000 tonnes of it around Broken Hill. Also locked in these tailings was the lead and silver the initial treatment had failed to recover.

After some expensive failures in trying to separate the zinc sulphide from the crushed rock by magnetic methods, mineral experts and others began to think of flotation processes.

Charles Vincent Potter was a Melbourne consulting brewer with a flair for invention, and had already invented a new nosebag for horses, a new stopper for bottles and a new way of making malt, when in 1900 he began to think about Broken Hill's zinc problems.

He crushed samples of Broken Hill ore, put them in a vat of hot water, then added sulphuric acid. Most of the particles of zinc and lead stuck to the bubbles created by the acid and rose to the surface, while the rest of the ore sank. By 1903, the Block 14 company had a small flotation plant working at Broken Hill, and it could recover up to 60 per cent of the zinc.

About the same time, Guillaume Delprat discovered a similar method. Dutch-born Delprat already had considerable mining experience when he came to Australia at the age of 42 in 1898 to become BHP's general manager at Broken Hill.

Part of the enormous tailings dumps at Broken Hill in 1904. These dumps held a fortune in zinc which for years eluded all attempts to extract it.

Delprat told his metallurgist, Carmichael, to boil up some salt cake and zinc tailings in an effort to get the zinc compounds to dissolve. Instead of dissolving they merely rose to the top on the bubbles caused by the salt. Carmichael declared the experiment a failure, but Delprat saw its significance, and BHP soon put the idea into practice.

Potter thought BHP's process was just too much like the one he had patented and sued them. After four years of litigation BHP won the case on the grounds that Potter's process lacked practical utility. Then BHP elected to use sulphuric acid instead of salt and the combined process became known as the Potter-Delprat.

One of the expert witnesses in the Potter-BHP litigation was Auguste de Bavay, brewer of Foster's Lager, who started his own experiments in separating the zinc sulphide from Broken Hill tailings. In July 1904 he mixed the crushed ore with oil and then dropped it on the surface of the vat. The zinc floated and the rest of the ore sank.

Two Melbourne financiers, Montague Cohen and W. L. Baillieu, joined de Bavay in forming the De Bavay Treatment Company to develop the process. Confident that the process would work eventually, Baillieu bought up enormous quantities of zinc tailings at a few shillings a tonne. After five years of trial and error the company had made the flotation process efficient and was reconstructed

Guillaume Delprat, general manager of BHP, who found one way to extract the zinc from the tailings.

as Amalgamated Zinc (De Bavay's) Limited, which built the world's largest flotation plant, at Broken Hill.

In the Potter-Delprat process, the sulphide ores attached to bubbles of gas made chemically, rose to the surface of the vat, but in 1905 workers at the central mill in Broken Hill realized that the result was the same when the bubbles were air. Air bubbles could be made far more economically by agitating water with paddles. A small London company, Minerals Separation Limited, made the same discovery about the same time.

A flotation plant at Broken Hill, 1915.

Modern flotation plant at Broken Hill.

Baillieu, who helped finance De Bavay's plant, was also deeply involved in the formation of Zinc Corporation Limited, along with American mining engineer and future President of the United States, Herbert Hoover, who visited Broken Hill in 1905 and bought millions of tonnes of zinc tailings for future treatment. Zinc Corporation Limited tried the Potter and Mineral Separation processes, then an English vacuum separation process, and finally settled for an improved version of the Minerals Separation method.

Even with the flotation methods separating the zinc sulphide from the pulverised host rock, there still remained the problem of separating the zinc sulphide from the residual lead sulphide missed in the original gravity separation. When the combined concentrate was smelted to free the zinc, the lead was lost.

The problem was solved in 1912 when F. J. Lyster, a carpenter who was working as a foreman for the Zinc Corporation, discovered a method of selective flotation. So too did T. M. Owen of the Junction North mine and Leslie Bradford of BHP soon after, although the processes were different. The discovery of selective flotation was due to a lucky coming together of all the right circumstances which Lyster happened to notice, investigate and patent. By September 1912 the Zinc Corporation had worked out how to perform selective flotation on a commercial scale. Various chemicals added to the water determined which particular mineral attached itself to the bubbles and floated to the top.

The idea of flotation in minerals processing spread from Broken Hill to other mines in Australia and all over the world. Chemists and metallurgists in such places as the United States and Germany were soon busily experimenting in attempts to improve the various processes.

But much of the most successful work was done in Broken Hill and Melbourne. Between 1902 and 1915 at least 40 men in 11 companies made worthwhile contributions to practical knowledge on the subject. By 1911, when the first successful froth flotation plant started in the United States, Broken Hill had floated about eight million tonnes of crushed rock and was the source of one fifth of the world's zinc concentrates. In his book about Australian mining, *The Rush That Never Ended*, Professor Geoffrey Blainey brackets the flotation process with the cyanide and Bessemer processes as one of the three great advances in metallurgy in the last thousand years.

18

Walter Hume's centrifugal painting machine led to the Hume-spun pipe

WALTER HUME was on his honeymoon on a boat from Adelaide to Kangaroo Island when the idea of a spun concrete pipe came to him. In concrete technology and practice, it was an innovation of international significance, for within the inventor's lifetime factories all over the world were making pipes by Hume's method.

Melbourne born Hume started work in 1886, when he was 12, providing the leg power for a treadle operated lathe, which earned him half-a-crown a week. From that beginning he went into the building trade, and learned something of plumbing, blacksmithing and plastering.

When he was 18, he and his elder brother Ernest started their own business, as carpenters and plasterers, but this lasted only until the collapse of the land boom. Penniless again, the Hume family (the two boys, their two sisters and their widowed mother) left Melbourne for the country, taking with them 'ten shillings and a fortnight's tucker'. They went to Moorooduc, on the Mornington Peninsula, where at least there were rabbits for the catching and firewood for the picking up. For two years they cut firewood at 30 shillings a railway truck, collected honey from the wild bees, and did odd jobs.

Then their uncle offered them a share-farming proposition near Kyneton, and this worked out quite well until the uncle died and the farm was sold. The family moved to Malmsbury and set up a buiness repairing farm equipment in an old hotel. It was here that Walter Hume conceived his first invention, a new type of fencing dropper.

In 1900, when Hume was 27, he and Robert felt they had progressed enough to start making the fencing droppers in Adelaide. They had a big contract for rabbit fencing in South Australia.

In Adelaide the Humes added a new line to their range, ornamental steel fencing. Painting this took every bit as long as making it, so Hume started looking for an easier way. Dipping each panel of fence was certainly easier, but excess paint spoiled its appearance.

Hume remembered that he had once seen honey extracted from its wax

Walter Hume attributed his success to 'persistent effort' and 'totally ignoring . . . orthodox theories'.

Removing the steel mould from a large concrete pipe.

combs by centrifugal force, and saw the possibility of removing excess paint from his fencing panels in much the same way. He designed a machine that would spin the painted panels, put it together himself in his back yard, and prepared to test it.

Hume had many doubts and fears about the new machine. He ran slow trials at first, and strengthened various weaknesses that these revealed. Once he felt it was ready for full speed testing, he dipped several lengths of fencing, put them in place ready for spinning, and gave his mother the honour of throwing the switch.

If the machine had flown to pieces no one would have been too surprised. Hume admitted later that turning on the machine for the first time was a terrifying episode, and that he turned it off sooner than he had intended.

In fact it had been a great success. The painting was done perfectly. The panels were completely covered with paint, yet there was not a surplus drop or blob anywhere. Hume strengthened and balanced the machine and put it into regular service. Three men could now do more painting in a morning than 20 boys had been able to do in a whole week.

From this particular success, Hume developed 'an encouragement to delve into mechanical contrivance'.

About this time there was great interest in pipes for irrigation in South Australia. On his honeymoon, late in 1909, it occurred to Hume that concrete could be made into pipes by centrifugal force in much the same way as the paint from his fencing panels was evenly spread on the inside of the shield around his machine.

Even before he had worked out how to put his idea into effect, Hume formed a public company to develop it. The prospectus for Humes' Patent Cementiron Syndicate Ltd, issued in February 1910, called for a capital of £3 000 in 60 shares of £50 each. Of these, 40 would go to the Humes, and 20 to subscribers, with the £50 payable £15 on application, the balance as required.

In the prospectus, Hume revealed only that the syndicate was 'being formed with the object of patenting and constructing a plant for an invention in connection with the combined use of cement and iron or steel.

'Owing to the novelty involved in the present invention particulars cannot be freely given until protection is obtained. The promoters admit that they are offering a pig in a poke, so to speak, but are, however, quite prepared to take the responsibility of reticence.

'The invention aims at practically securing a monopoly in some of the largest and most needed artificial necessities on earth.

'The fact that Messrs Hume Bros have made a success of their Rolled Steel Patents, and, as a firm, have an honest record since their start 18 years ago, should prompt confidence.

'The approximate cost of bringing the scheme up to a proved and world patented monopoly is estimated at about £1 000. As soon as this is accomplished there should be no great difficulty in forming subsidiary companies, selling sectional Rights, granting Royalties or promoting companies for undertaking the manufacture.'

Hume freely conceded at the time that his idea might fail, in which case investors in his company would lose their money. But if he succeeded, he said, a £50 share might soon be worth £15 000. He soon had the 20 subscribers he needed.

Secrecy was vital, so Hume began experimenting on Saturday mornings, when none of his employees were in the workshop. He put some mortar inside a steel pipe, and set it spinning in a lathe. His early tests proved that surplus water would separate from the concrete, and that the concrete would pack tight upon the inner surface of the pipe.

Once he had lodged his patent application, Hume felt he could let the secret out enough to have a larger spinning machine built. It was so arranged that the mould could be spun either vertically or horizontally. Vertical spinning seemed to give the best results, but the concrete always tended to finish up as an inverted cone, and stick so tightly to the mould that it could not be separated.

After weeks of disappointment, Hume was beginning to feel that maybe the idea just would not work with concrete. He decided to try out its possibilities for making cast iron pipes. He did a test run with molten lead, and decided that to make metal pipes by a centrifugal process would require the mould to be spun much faster. He built a new machine that drove the mould by friction at a very high speed.

Once he had built his new machine, in November 1910, Hume decided to try it first with concrete. Almost by accident he had discovered the secret that had eluded him, for with some modifications the new machine made concrete pipes successfully.

A seven feet (2.1 m) diameter Hume spun pipe.

A consignment of 152 Hume pipes leaves the Brisbane works for the Coolangatta drainage scheme.

To get acceptance for his new pipe, Hume called on S. MacIntosh, Director of Irrigation in South Australia, who after examining the pipe exclaimed, 'Well, I'm damned, I've been all around the world looking for it, and here the very thing I want is on my doorstep'.

Hume gradually perfected his process so cement pipes could be made in commercial quantities and in a range of sizes. The 1920 prospectus for a new company, Hume Pipe Company (Australia) Limited, explained that in a typical Hume pipe plant, a cylindrical mould with movable cast-iron ends revolved on a friction drive. Mortar thrown in at the ends soon distributed itself equally around the inner perimeter of the mould. Centrifugal force condensed the mortar to a high density. Superfluous water was soon expelled and since there was no inner mould to impede its escape, a finished pipe could be produced in a few minutes.

When reinforced pipes were required they were made by inserting a machine woven cage of slightly smaller diameter than the mould. The mortar soon embedded the reinforcement in the wall of the pipe. Pipes were usually made in lengths from 0·9 to 3·6 metres, and with internal diameters of 50 mm to 2·4 metres.

The Hume process was adapted to other products, such as this Melbourne telephone booth.

Pipes with diameters up to 152 mm were made four at a time on a single machine and the four finished and rolled off in their cases within six minutes. A couple of 760 mm diameter pipes 2·4 metres long could be made in 12 minutes. The centrifugal action was so effective in compressing, drying and finishing the pipes that even a thumb pressed into them with all a man's weight would not mark them.

Spun pipes were placed in their cases in a steaming chamber for four to six hours, then taken into the open, stripped of their cases and rolled off as finished pipes. All this was done within 12 to 18 hours, but the pipes were usually left for from three to 21 days, according to their intended use.

Hume had revolutionized the making of concrete pipes in more ways than one. Pipes previously cast in a combination inner and outer mould had not been able to stand much pressure. Pipes made by the Hume method withstood tests of up to 400 pounds per square inch. The centrifugal action made them particularly dense. Since all the water was expelled the pipes were not porous, for no air spaces were left where water had dried out.

The equipment needed was simple, and could be set up near the site of any major project calling for a large supply of pipe. Even in 1920 Hume was able to tell English investors that a mobile plant could be had to make pipes with an internal diameter of 2·4 metres, a length of 3·6 metres, and walls 76 mm thick.

The centrifugal process proved to be adaptable to a wide range of hollow concrete products: columns for buildings, cylindrical, tapered or ornamental; poles for telegraph, telephone and electric wires; fencing posts of all shapes and sizes; tiles, ridging, guttering, chimneys, building blocks, hollow bricks from ordinary building sizes up to 2·4 metre cubes; factory chimneys with either single walls or double walls filled between with concrete; ventilation pipes; gas mains, oil mains, sewer pipes, water tanks, incinerators, silos, troughs and so on.

Humes soon had factories in all States and the process spread fairly rapidly overseas, starting in New Zealand in 1916, and South Africa in 1917. By 1920, factories using the process were also to be found in the United States, England, Japan and Germany. During the 1920s, India, Hawaii, the Philippines, Brazil and others joined the list. Hume set up a factory himself in Singapore in 1922.

Hume's company developed its expertise in pipes to include not only concrete ones, but also steel pipes, and a combination of the two, concrete lined steel pipes. In 1923 Walter Hume patented 'a method for curving sheet metal to form pipes and other tubular products' and an 'improved automatic arc welding machine'. He had accidentally discovered the idea of lining steel pipes with concrete when he tried to make his first concrete pipes only to find that they would not come out of the mould. He solved the problem then by oiling the inside of the mould before spinning the concrete. But he also realised the potential of concrete-lined steel pipes, and they became an important Hume product. Nowadays the company he founded also lines pipes of various kinds with plastic, a process that can be applied to existing pipelines.

Before he died in 1943 Hume wrote of his highly inventive life: 'My success in life has been due perhaps more than anything to my persistence in carrying out my purpose. My many inventions have been brought to successful conclusions in nearly every instance by a most persistent effort, and what scientific faculties I have seem to be centred in the power to discern and note simple and important things and put constructions on them as please my scientific sense, totally ignoring anything in the way of orthodox theories.'

19

John Dethridge's wheels and weirs

DRIVE AROUND THE irrigation areas of Victoria and you are bound to notice many mini-water-wheels turning slowly beside the road, as water in the irrigation channels flows beneath them. You are not so likely to notice, unless you look very closely, the unique construction of the weirs across the Murray River at Mildura and Torrumbarry.

Both are the work of John Stewart Dethridge, one time Commissioner of Victoria's State Rivers and Water Supply Commission.

The Dethridge Wheel is rather like a scale model of the sort of waterwheel once used to power flour mills and mining machinery. But the only thing it powers is the geared revolution counter linked to the axle of the wheel. Each revolution of the wheel allows a known volume of water to pass under it, and the meter is

John Dethridge.

A Dethridge wheel.

arranged to register how much water each farmer has drawn from the channel. A sluice gate near each wheel enables the farmer to control the rate of flow of water.

The Dethridge weirs across the Murray River at Mildura and Torrumbarry are a kind unknown in other parts of the world. They consist of steel trestles on wheels which are run across the river on concrete and steel tracks. In normal use they hold the water about three or four metres above summer level, so that it continues to flow into the irrigation channels. But upon the approach of floods, they are pulled out of the river altogether. Painting and maintenance can be done easily at such times.

20

Lance de Mole's tank was just before its time

In 1912, two years before World War I began, North Adelaide engineer Lance de Mole sent the War Office in London a proposal for a 'chain-rail vehicle which could be easily steered and carry heavy loads over rough ground and trenches'.

The War Office rejected the invention because they were 'not further experimenting with chain rail vehicles'.

This rejection may have stemmed from the sort of incompetence that some people find so easy to attribute to such bodies as the War Office, and not always without reason. Lance de Mole was a colonial, and in 1912 colonials were not usually expected to be able to show their betters in the Mother Country how things ought to be done. The idea of powered vehicles sheathed in iron lumbering across the countryside probably did not fit the dashing image of the cavalry officers of the time.

But then the reason for the lack of interest in de Mole's tank may simply have been that the War Office could see no likely use for it. All their recent experience was with the Boer War, and dashing about after unruly tribesmen in India. Trench warfare had been known before, but not on the scale that was to occur within the next few years in France. It was the need for some means to penetrate enemy territory and break the stalemate on the western front that brought the tank into its own. About the same time that the British War Office was turning down de Mole's tank, the Austro-Hungarian and German general staffs were turning down similar ideas.

In the case of the tank, necessity as usual was the mother of invention. When the War Office discovered the necessity for some machine such as the tank, they put people to work inventing it. Unfortunately, de Mole's idea was by then gathering dust in the War Office files.

The British tanks that appeared in France in 1916 were invented by Major Wilson and Sir William Tritton, who got a £15 000 reward from the British Government for their work.

The British Royal Commission on Awards to Inventors agreed that de Mole 'had made, and reduced to practical physical shape, as far back as the year 1912,

Model of de Mole's tank at the Australian War Memorial, Canberra.

a brilliant invention which anticipated, and in some respects surpassed that actually put into use in the year 1916; and this invention was in fact communicated at the time to the proper Government Department.'

The tank designed by de Mole could be steered by screwing the tracks to one side or the other, so they formed a curve. The four bogies under the vehicle could move sideways to follow the curve of the chain rail. The whole vehicle could turn in a radius of about 20 metres. It was about eight metres long and could cross a trench nearly four metres wide. The Australian War Memorial in Canberra has a one-eighth scale model of the tank which was built under de Mole's direction.

The British Royal Commission regretted that it could not recommend any reward to de Mole and that there was no direct connection between his invention and the tanks the British Army first put into the field in 1916. The Commission conceded that 'it was this claimant's misfortune and not his fault that his invention was in advance of his time and failed to be appreciated and was put aside because the occasion for its use had not arisen'.

21

Headlie Taylor taught himself engineering so he could start inventing

IN 1920, many Australian wheat farmers were looking forward to a bumper crop. Good rains in the eastern States had produced crops that were 'over the fences' and laden with grain, particularly in north-west New South Wales.

Suddenly, all their hopes seemed to be dashed. Violent storms in November lashed the heavy crops to the ground, flattening and tangling them to such a degree that there seemed no hope of a successful harvest.

This was the great opportunity Headlie Shepherd Taylor had been waiting for. The McKay factory at Sunshine worked day and night all through November and December that year, turning out over 1 000 new style Sunshine Headers, each fitted with a special Headlie crop lifter, a set of wooden fingers to lift the crop on to the comb.

Astonished farmers got 30 bushels of wheat to the acre from crops they thought had been completely ruined. Never before had such downed and tangled crops been harvested without heavy loss.

Taylor's triumph was complete. But the sudden appearance of his miracle header on the wheatfields was the result of long and painstaking work. Born at Henty, in southern New South Wales, in 1883, Taylor had grown up without much education, but he had become familiar with farm machinery, and had become convinced that he could improve on McKay's Stripper Harvester.

He agreed that it worked well enough in standing crops, but he felt that its efforts to rescue the grain from down and tangled crops resulted in considerable waste. He decided to devise a machine that would overcome this problem.

By January 1911 he had his general ideas worked out, and decided to build a working model. He set about a conscious plan of book buying and study. He bought books about mechanical drawing, books about pattern making, books about moulding, in fact any book he thought might help him. The engineering knowledge he got from these books, plus his own natural commonsense, was all he had to guide him.

He began by enlarging his farm workshop, so he could install a lathe and a power drill, and allow room for the machine and working space around it.

Taylor with one of his original harvesters.

Taylor accepted that his progress would be slow, with long hours of tedious work. He thought it might be two or three years before he would prove his success or failure. He had confidence that he could do the job, but he was worried about finance for it.

He worked day and night to have his first machine ready for an initial trial in the 1911-12 crop, and managed to finish it early in the new year. By then most of the crops had been harvested, but there was a small paddock of standing crop close to the workshop which he had kept just for the test.

At his first attempt, Taylor took nearly half a day to get once around the paddock. He had frequent stoppages and fresh attempts to get the machine going, but the comb simply would not go into the crop. It knocked down more heads than it gathered.

Disappointed, Taylor took his machine back to the workshop and wondered what he might do to overcome the problem. He could see that there was no hope of a second trial that season, so he wisely gave up the task and took a holiday. He was lucky in that his two brothers had agreed to do all the work on the farm with the exception of shearing while he concentrated on his invention.

After three weeks of rest, Taylor came back to the job with renewed vigour. He redesigned the whole machine, retaining only the wheels and axle of the original one. This time he hired a helper, and had the new machine ready in good time to try it out on the 1912-13 crop. With a few minor adjustments the header worked well, and took off about 81 hectares of grain, some of it fairly heavy crop and badly down and tangled.

Taylor took out his first patents in 1913, and was granted more claims for novel features than had ever been granted for a harvesting machine in Australia previously.

He now set out to build a machine that would be unbeatable, and he planned to show it at the Henty Show in 1914. He managed to finish it with two days to spare. It worked very well in a ripened wheat field.

After a few days of experimenting, Headlie Taylor challenged his brother Horace to a match with his harvester. The match would be decided on the best average yield of wheat per hectare. Horace readily took up the challenge but was decisively beaten.

Encouraged by the result of this match, Taylor decided to show his new machine to as many farmers and machinery manufacturers as he could muster.

Engine-functioned but horse-drawn, this rice header was designed at Sunshine by Headlie Taylor. The engine permitted efficient harvesting regardless of ground speed.

1965 Sunshine header was self-propelled, had a 4 or 5 metre comb front, 15 speeds in each of three forward gears.

He placed an advertisement in the local newspaper, and wrote to manufacturers. He had a good turn up at his demonstration, which was so successful that several influential farmers offered to finance a company with a capital of up to £59 000 to build the new machine, provided that Taylor would supervise the work.

Taylor considered this idea, but decided it would involve him in too long a commitment. He chose instead to create a competitive desire for his invention among Australian machinery manufacturers. He decided the best way to do this was to get the machine on the market.

He found it easy enough to get orders, but a little harder to get the machines built, since this needed finance. An Albury man helped by promising to advance as much money as he needed, and he soon arranged with a machinery manufacturer in Melbourne to build five machines.

Work began in April 1915 on making all the patterns and tools that would be needed to make the Headlie Header in quantity. Although Taylor had to pay for all five machines only three were ready for the 1915-16 harvest. Because of the setting-up expenses these three machines cost Taylor £1 500 each, but to be competitive with other harvesters he had to sell them for £150 each. So he was well out of pocket on the deal.

Taylor now arranged another demonstration, in a crop that was tall and heavy, on a Henty farm. And this time he wrote direct to the biggest agricultural machinery maker in Australia, Hugh McKay. A few days later he heard from McKay that he would have to miss the scheduled trial, but that he would be along a few days later. The planned trial went well, but the special trial for McKay went even better.

The famous inventor of the Sunshine Harvester arrived as promised, and with pardonable trepidation Taylor began to drive his header through the crop, with McKay sometimes walking behind, the better to observe the action of the machine.

McKay agreed that Taylor's harvester was very simple, very light and that it had many novel features. But what, he asked, made Taylor think it was better than his own Sunshine Harvester. On being told that it could handle a heavy, lodged crop and get almost all the grain from it, McKay said that if such were the case he was prepared to negotiate for the patent rights.

McKay must have been convinced, for on 4 March 1916 he and Taylor signed an agreement which provided that Taylor was to come to Melbourne and supervise the building of the header for at least two years, and make further alterations he thought advisable after watching the operating results of the three machines already built.

The first programme at Sunshine was for six machines, to be ready for the 1916-17 harvest. Fresh templates were made and Taylor, with his uncertainty and doubts about finance behind him, busied himself with the task of getting his invention into mass production.

Yet even Taylor was amazed by the success of those six machines, called Taylor Headers. Their performance in the 1916-17 harvest led to a flow of orders that called for more factory space and plant. The following year there were even more orders, and more factory space and plant was needed. This was the pattern until the crowning success of the 1920-21 crop, which placed Taylor's machine, now called the Sunshine Header, at the forefront of farmer favour in harvesting machines.

The main difference between Taylor's machine and McKay's was that Taylor's cut the heads from the stalks instead of beating them off. It had a system for quickly but gently removing the heads from the openings of the comb the moment they were cut off. Two spirals ran the full length of the comb, the front one gathering up the incoming heads, the two together providing a conveyor to take the heads to the elevator and then to the threshing drum.

The comb stayed level as it was adjusted from its lowest to its highest point, which meant it gathered all the grain at every level. In previous harvesters grain had tended to fall off the points of the comb when it was set low, the heads to be pushed away when it was set high.

Two quite small fans attached one to each end of the threshing drum and enclosed by part of the drum case in the Taylor header replaced the large fan case and fan used in previous models, This made the new machine more compact. The main drives were distributed to avoid undue wear on any part, which saved farmers money on spare parts.

Taylor's two years with McKay's firm developed into a stay of 38 years, until 1954, when he retired. By the time of his death in 1957 he had become one of the most prolific inventors of farm machinery in Australia.

He adapted his header to suit various crops and environments. To cope with the boggy ground of the Murrumbidgee Irrigation Area rice crops, which made ground drive machinery useless, he developed a special header with its own engine to work the machinery. He invented the fully self-propelled auto-header in 1924, an idea which caught on better in America than it did in Australia, where many farmers still preferred to use their farm tractor. He also produced pasture renovating machines and bagging machines.

22

George Nicholas made Australia's first aspirin

USING KEROSENE TINS and kitchen utensils borrowed from his wife, Melbourne pharmacist George Nicholas set out in 1915 to react salicylic acid, a white powder, with acetic anydride, an acrid smelling liquid.

Pungent smells filled his pharmacy. Toxic fumes made him ill. He lost weight. At one stage his sight even failed him, but fortunately a spell away from his putrid pots enabled it to recover.

The 31-year-old Nicholas was trying to make the first acetyl salicylic acid,

George Nicholas at his desk in the Aspro factory, City Road, Melbourne, about 1920.

The Melbourne chemist shop where George Nicholas worked out for the first time in Australia how to make aspirin.

or aspirin, ever made in Australia. Until 1914 all Australia's aspirin came from Germany, and supplies had been cut off by the outbreak of World War I.

Knowing the chemical reaction for making aspirin was one thing. Getting it to work on a commercial scale was something else. Nicholas was determined to find the secret.

Eventually Nicholas did make aspirin of a sort. He heated the mixture for hours and let it cool slowly. As he watched, white flakes formed and joined together. But this aspirin was impure, soggy and mixed with malodorous liquid.

Nicholas sought the help of freelance industrial experimenter Henry Woolf Shmith in his attempts to purify the drug. As their experiments progressed, Nicholas often cat-napped through the night on a couch in the lab to be near the chemical reactions at crucial moments.

Nicholas and Shmith managed to drain the obnoxious liquid away, but the remaining aspirin still had a vinegar taste, a brownish pink colour and a vile smell. To get rid of these faults, Nicholas and Shmith tried dissolving the aspirin in ether and re-crystallising it. This worked well enough, but it was dangerous. One day a tin of ether caught fire and Nicholas managed to put it out only by sitting on it.

The tablet making machine on which Nicholas made his first aspirin.

After many weeks of experimenting Nicholas and Shmith had a batch of pure aspirin. On 12 June 1915 they applied under wartime legislation to take over the trade name 'Aspirin' from Bayer, a German firm.

Officialdom ignored their application until Labor politician Frank Anstey prodded Labor Attorney-General Billy Hughes into having the government analyst test the Australian made aspirin. Hughes on 17 September 1915 announced that it was purer than the German product. He granted Nicholas and Shmith a licence to make and sell Aspirin in Australia.

The two men's Aspirin venture was soon losing so much money that Shmith pulled out, leaving the project to George Nicholas and his brother Alfred. In April 1917 they registered the name Aspro. In time the business prospered and eventually grew into Nicholas International Limited, an Australian-based multinational corporation marketing Aspro all round the world, along with a large range of pharmaceuticals and cosmetics.

George Nicholas made his last official appearance in 1957, when the firm opened a new modern factory in the Melbourne suburb of Chadstone. He spoke of his start in the little Melbourne pharmacy. 'Conditions were really awful,' he said, 'but it was the enthusiasm that kept us going . . . We couldn't have imagined that we were planting the seed that has germinated into this.' He was 76 when he died in 1960.

23

Periscope rifles saved Anzac lives at Gallipoli

Some of the most savage fighting of World War I took place on the Gallipoli Peninsula, in Turkey, where Australian and New Zealand troops landed on the first Anzac Day, 25 April 1915.

The Australians quickly dug in and the campaign became stalemated in trench warfare, with the trenches in places only five to seven metres apart. To look over the top in such circumstances meant almost certain death.

Resourceful Australian troops soon proposed a solution. On 19 May 1915 Major Thomas Blamey, the future Field Marshal, saw two men engrossed with a framework of broken box-wood and wire, attached to a rifle. On asking what it was Blamey was told 'an arrangement you can hit with without being hit'. It was a periscope arranged for firing a rifle. The upper glass looked along the sights, while the sniper looked into the lower one.

Blamey arranged for the inventor, Lance-Corporal Beech, of Sydney, to go to headquarters to demonstrate the idea, with the result that a factory to make periscopes for rifles was set up on the beach. Tested at many spots, it proved an accurate and deadly weapon at ranges of 200 and 300 metres. It was constantly used by Anzacs wherever the trenches were close.

Another invention based on a rifle helped the Anzacs evacuate Gallipoli some months later. British generals estimated that half the force would be lost in any attempt to withdraw, the Turks being almost certain to discover what was going on because of the closeness of the trenches.

Among the devices used to deceive them were self-firing rifles. One version of this had two tins, the upper one filled with water that gradually dripped through a hole into the second. When the lower tin became heavy enough it overbalanced and fired the rifle by pulling a string attached to the trigger. A variation on this was a candle that burned through a string and released a weight attached to the trigger. Such devices provided sporadic firing long after thousands of men had crept down to the beaches and got away. The Turks were so deceived that of the 80 000 men withdrawn from the Gallipoli Peninsula only about half a dozen were lost.

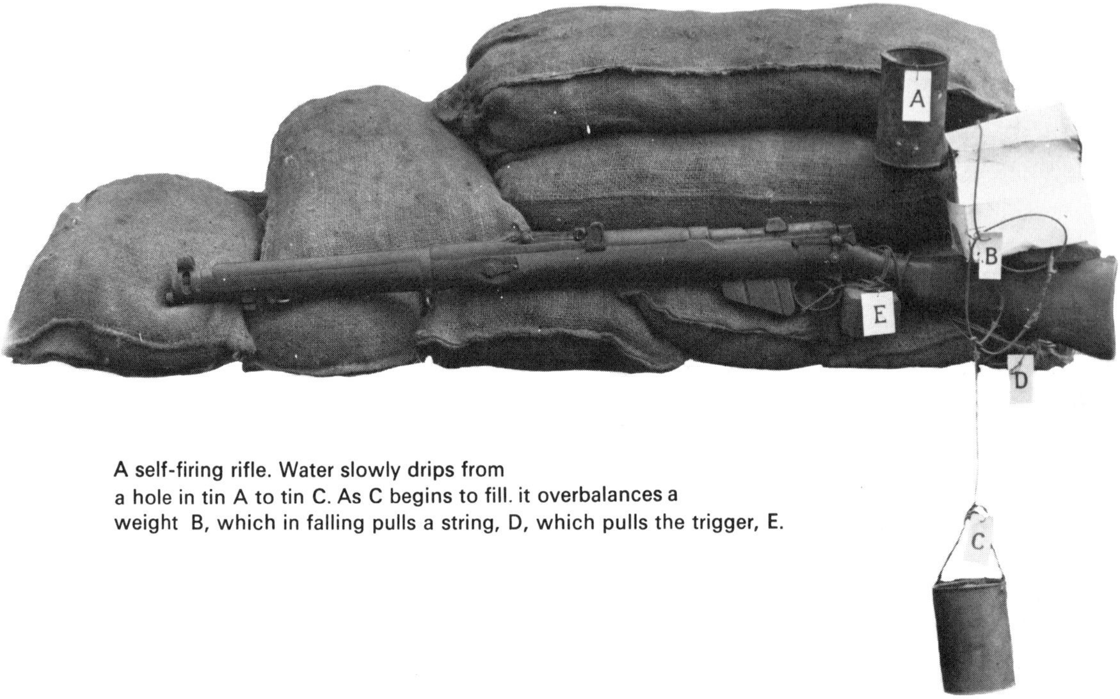

A self-firing rifle. Water slowly drips from a hole in tin A to tin C. As C begins to fill. it overbalances a weight B, which in falling pulls a string, D, which pulls the trigger, E.

An Anzac at Gallipoli uses a periscope rifle, while a comrade spots for him through a periscope.

Another Australian soldier of World War I who showed a talent for invention was Corporal A. R. Muirhead, who served in France. He spent six months working out a device for reloading the drums of Lewis guns, only to find on taking his model to London when he was on leave that somebody else had beaten him to exactly the same idea.

The army officers who had to tell Muirhead the bad news tried to soften his disappointment by offering him another task that called for an inventive mind. If he could work out some way to avoid the constant jamming of machine guns, they said, they would be more than glad to hear from him. All the great nations of the world had been working on the problem for years, but nobody had found a solution.

In tossing this problem to Muirhead, the officers may have already become aware of the knack Australian soldiers seemed to have for wartime invention. There was an uncanny soundness about many suggestions from Australian troops. In many cases, when the stresses and constructional details of their ideas were worked out mechanically, they were found to be almost identical with the original rough drawings.

Muirhead said he would like to have a week or two to think of the problem of the jamming machine guns. But he worked all night and turned up at the research station next morning with a model made of fencing wire, clothes pegs and a picture hook. And it worked.

Within two weeks a complete model had been built, but then Muirhead said he wanted to scrap the idea completely and alter the whole practice of machine-gun belt construction. He proposed that belts should be made of paper and used only once. The whole idea was so obviously right that prototypes were made at once and they acted perfectly.

Muirhead's method was adopted, and within six weeks of its invention the new style of belt was being made.

24

Cliff Howard built a business on his rotary hoe

ALTHOUGH CLIFF HOWARD was only 16 when in 1912 he first tried his idea that some kind of rotary movement could be used to replace traditional ploughing, he persisted with developing the idea for the rest of his life, and built a thriving business on it.

With his farmer father's encouragement, young Howard rigged a drive from the family tractor to the shaft of a one way disc cultivator with the discs notched so they would grip the ground. He took gear chains, sprockets and cog wheels from a derelict reaper and binder for his transmission.

This 1912 test, on his father's farm at Gilgandra, New South Wales, showed that ground could be tilled without the wheel spin and soil packing that occurred in normal ploughing. It also showed that heavy weed growth could be cut through and mixed with the soil.

But the fast spinning discs threw the soil sideways far too much, and Howard knew he would have to think of some new tool if his idea of rotary tillage was to work. He tried many ideas until he finally developed an L-shaped blade mounted on flanges fixed to a small diameter rotor.

Once Howard had a model that worked effectively he went into business with a fellow apprentice, Everard McCleary, to make rotary hoes. At first they thought in terms of small machines. The prototype used the engine from Howard's motor-bike. But they found little demand for such small models.

They then turned their attention to a machine big enough to interest wheat farmers. They developed one that could cultivate at the same rate as an eight to twelve-horse ploughing rig, and at the same time sow seed through the attached seed drills. The chassis of this machine was so designed that the rotary hoe could easily be taken off and a combined reaper and thresher mounted in its place. The machine was therefore suitable for the wheat farmer's entire annual cycle.

McCleary was killed in World War I. Howard had been injured in a motor-bike accident and was unfit for active service. However, he went to England and worked on munitions and aircraft engines.

After the war Howard stayed in England and tried to get agricultural implement firms to develop his idea. Some seemed interested, but would not spend any money on the project. So Howard in 1919 came back to Australia and began to build a new rotary hoe himself, using old sulky springs for the blades. The machine, which he towed behind an eight-tonne Buffalo-Pitt steam engine, had

Cliff Howard in 1958.

five sets of one metre rotors, so that it cultivated a strip of ground five metres wide. When tested at Gilgandra it performed perfectly.

Howard formed a syndicate among his friends to finance his venture, which he registered in March 1922 as Austral Auto Cultivators Pty Ltd. He soon sold his first cultivator for £1 500. The firm operated from the Moss Vale plant where Howard had served his apprenticeship.

Cliff Howard driving an early Rotovator in 1922.

For a time lack of money limited development to this particular unit. But as finances started to ease a little, Howard designed and built a rotovator that fitted a Fordson tractor, a smaller unit that could be built and sold quickly, thus permitting a faster turnover of working capital.

Hand controlled machines followed.

By the end of 1925 all available labour at Moss Vale was fully occupied. Moreover, transport of raw materials to Moss Vale and transport of finished

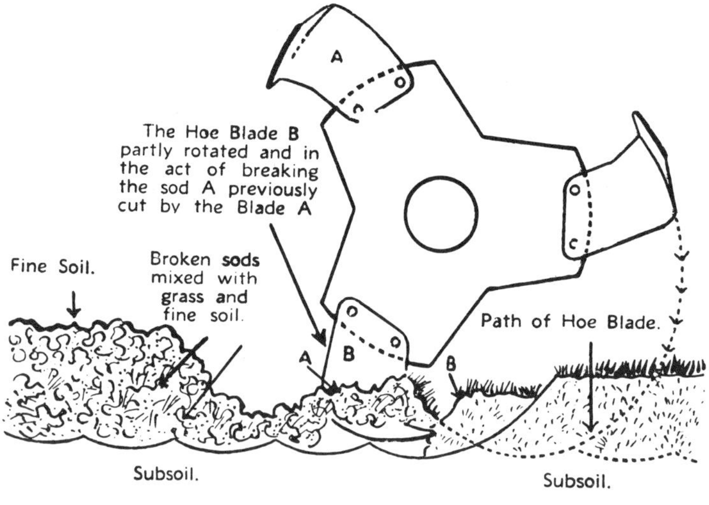

The Howard Rotary Hoe patented method of tilling the soil. Note the rippled surface on the subsoil. **With this method hard pan is impossible.** It also assists in holding the loose soil from washing away with heavy rains.

Diagram from a Howard advertising brochure showing how the rotary hoe tilled the soil.

products was costly. So Howard bought land in the Sydney suburb of Northmead and in 1927 moved his plant there. In the same year he produced his first DH22 tractor, designed to work with his rotary hoes.

To expand his sales potential, Howard designed special models for particular jobs, some for Queensland sugar cane farms, some for clearing up rough country and others for dealing with weeds and prickly pear.

Like most Australian businesses, Cliff Howard's firm had a hard time in the depression, particularly as it was just getting established when the economy started to turn down. But once the Depression was over business boomed, and Howard went to England, took over a company he had licensed there to make Howard rotary hoes, and made it the basis for the world-wide manufacture of them. He was awarded the Order of the British Empire a year before his death in 1971.

25

Alf Traeger's pedal wireless eased outback isolation

As settlers pushed into the outback of Australia in the latter part of the nineteenth century, they cut themselves off completely from most of the trappings of civilization.

On isolated sheep and cattle stations, a small group of settlers, perhaps just a man and his wife and children, might go for a year at a time without any contact with the outside world. It probably came, when it did, in the form of a teamster or bullock driver who brought in the annual stores.

At best this meant loneliness and monotony. At worst it meant death, for there was no way of getting medical help if it were needed in an emergency. If a stockman was thrown from his horse and badly injured, his family or friends could only make him as comfortable as possible and hope for the best. Sudden illness often meant sudden death.

When John Flynn began his work as a missionary in outback Australia he began to appreciate just how serious this problem was. Born in Adelaide in 1880, Flynn had worked as a schoolteacher briefly before becoming a Presbyterian minister. He worked for some years in remote areas of South Australia before he persuaded the church to give him a broader area for his missionary work, as a result of which the Australian Inland Mission was established.

Always a practical missionary, Flynn saw that improved access to medical advice and attention was badly needed in the outback. In the 1920s he conceived the idea of a Flying Doctor, one who came by aeroplane to the spot where he was needed. But the problem was to find some means of summoning the doctor. In the outback in the 1920s telephones were comparatively unknown, as they still are.

Flynn soon realized that wireless was the answer, but his specification for a suitable set was rather too advanced for the time. It had to be able to receive and transmit for up to 480 km, be easy to handle by non-experts, weigh less than 22 kg, and cost £50 or less. It had to be foolproof, and easy to maintain and install, for there were no radio technicians in the outback.

In 1925, George Towns, an invalid soldier, and Harry Kauper, chief engineer of radio station 5CL Adelaide, built an experimental set which sent messages for

480 km when tested in the country around the Birdsville track. The set was clumsy and temperamental, but it was the basis of the future successful pedal wireless.

In 1926 Harry Kauper introduced Flynn to Alfred Traeger, a young Adelaide electrical engineer. Traeger and Flynn set up three experimental stations, a base at Alice Springs, and outstations at Hermannsburg Mission, 130 km to the west, and Arltunga, 110 km east. The mother station had a power of 50 watts and was powered by batteries.

These experiments showed promise that radio could do the job. But Traeger was not happy about using batteries, which were hard to transport over rough

Alf Traeger with an early pedal wireless.

bush tracks. He was concerned too that their acid was likely to spill or evaporate, or be drunk by children in station homesteads.

Back in Adelaide, Traeger fixed a small generator to an emery wheel and turned it by hand. This worked, but left the operator with only one hand free to send morse. To overcome this problem, Traeger screwed the apparatus to the floor, fitted bicycle pedals, and thus invented the famous pedal wireless. Anybody at all could generate 20 watts of power at 300 to 400 volts.

There was still the problem that any irregularity in the flow of power produced frequency changes in the radio. This could be avoided only if the operator could pedal very steadily. Traeger overcame this by importing crystal frequency stabilizers from the United States. And he increased the range of the set by using short wave instead of medium wave frequencies.

Towards the end of 1928 Traeger was turning out workable pedal wireless sets. The first of them was installed at Augustus Downs, 320 km north of Cloncurry in Queensland, in June 1929. Other pedal radio sets were soon placed at Gregory Downs, Lorraine, Mornington Island, Corinda, Birdsville and Boulia, outposts that were up to 640 km from the base at Cloncurry, where Australia's first Flying Doctor service had started in May 1928.

Traeger had to stay several weeks at each place he installed a pedal wireless, long enough to teach the owners morse code, for although the base could use voice transmissions to the outposts, they could only transmit to the base in morse. To overcome the problems of garbled messages, Traeger now invented a sort of morse typewriter. When the operator pressed a key, the machine sent a perfect morse signal on the pedal transmitter. The method was slow but sure.

Traeger continued to work on his pedal radio and by the early 1930s had produced a transceiver which could send voice in both directions. By 1946 he was building sets in which a battery operated vibrator had supplanted the famous pedals.

Although its primary purpose was for calling the doctor, the outback radio came in handy for many other purposes. Outback wives soon learned that they could use the radio to talk to other outback wives, and the daily 'galah session' became an established institution. More mundane business such as ordering supplies to be sent out with the mailman could also be conducted easily on the radio.

Later, the outback radios did extra daily duty for various schools of the air, in which a teacher could talk to a class spread over a vast area. After the first school of the air began operating in Alice Springs in 1950, others followed at Broken Hill in New South Wales; Ceduna and Port Augusta in South Australia; Charleville, Mt Isa and Charters Towers in Queensland; and Meekatharra, Derby, Kalgoorlie and Port Hedland in Western Australia.

26

Falkiner's sugar cane harvester worked well overseas, but not in Australia

WHEN MEMBERS OF the Australian Sugar Producers Association met in the Queensland town of Bundaberg in 1925, they were more than a little interested in demonstrations of a sugar cane harvester which had been invented by R. S. Falkiner.

Cutting cane by hand was one of the most back-breaking jobs in agriculture, yet every stick of cane throughout the world was harvested that way. Anything that promised to put an end to such labour was bound to attract attention.

There was an economic factor too. Manual cutters were expensive, not always easy to recruit, and inclined to be hard to manage. A mechanical harvester a farmer could drive himself promised freedom from many of the costs and problems that came with the cane-cutting season.

Falkiner, who was born in 1877 into the Falkiner family of stud merino sheep breeders in the Riverina, had developed a distinctly mechanical bent as a young man. He had first become interested in the problem of mechanically cutting and topping sugar cane in 1908, even though at that time he had never seen it.

Falkiner had taken over a cane-harvesting machine first developed by a Bundaberg sugar farmer named Hurrey, and financed the building of several experimental machines. The Hurrey machines could cut cane quite successfully at the thick ground end, but could not handle the more difficult problem of cutting the leaves off the top and disposing of them.

One problem was that stalks of cane in a row were different lengths and the machine could not adjust to them. Falknier developed a new system in which the sticks of cane were bundled, turned right over, and stood on their top end. This butted the heads of the sticks together on a baffle-board, where a knife decapitated them. This worked well enough, but too slowly to be a commercial success. Five or six men could cut just as quickly and just as cheaply as the machine.

To overcome this, Falkiner developed a system of rollers which threw each stick of cane head first against a baffle-board where a pair of knives decapitated it. A powerful air blast separated each stick of cane from the leaves and trash. The cane dropped into a bin on the side of the harvester, from where it could be transferred to other means of transport.

Falkiner cane harvester, side view, showing method of discharging cane.

Falkiner cane harvester, front view with operator in position.

The Falkiner cane-harvester worked well in straight, upright cane which had been burned. But much Queensland cane grew so heavy that it became twisted and lay at crazy angles, and the Falkiner Harvester was never successful in that sort of cane.

Falkiner took his machine to America where, with the help of American engineers, he eliminated most of its minor defects and evolved a much lighter and considerably less expensive model. The machine came to be widely used in Cuba and the southern United States.

27

Decades of patient work led to process for making paper from eucalypts

OVERSEAS EXPERTS SAID that making paper from Australian timber was impossible. Paper had always been made from softwood timbers with long fibres and Australia's eucalypts were hardwoods, with short fibres. But thanks to years of patient research by such men as Boas, Benjamin and Somerville, making paper from eucalypts is now an everyday occurrence.

The Tasmanian Government first brought an American chemical engineer to the State in 1914, in the hope that he might find a way, but he soon admitted defeat.

Still, the idea persisted. In Western Australia in 1918 the man in charge of State Forests, C. E. Lane-Poole, discussed it with I. H. Boas, who was then in charge of the Chemistry Department of Perth Technical College. Boas agreed to conduct some experiments on wood from karri saplings.

'We were then, fortunately as it turned out, almost completely ignorant of the techniques of pulping and paper making,' Boas's assistant Lou Benjamin wrote forty years later. 'I say fortunately because it led Boas away from making those initial experiments along the lines of the then current pulping practice.'

Boas and Benjamin set to work 'cooking' karri and jarrah, and soon convinced themselves that paper certainly could be made from Australian timbers. It was just a matter of finding how.

Boas went overseas to study forest products research and came back so full of enthusiasm that he persuaded the newspapers of Western Australia to provide £600 for a model paper making machine and other equipment for his experiments. But he failed to persuade the Commonwealth Government to provide a laboratory on the scale he felt necessary and gave up his experiments in favour of a job with private industry in Victoria.

Benjamin carried on and soon had samples of white paper made on the model from pink karri and even from red jarrah. He persuaded various State governments to provide a fund of £1 800 a year, which in those days was enough to pay five research workers. John Somerville became the group's chemist, and started a 25-year association with Benjamin. They could not afford a chipping machine, so one of the assistants chipped up the samples with a tomahawk.

Lou Benjamin, research pioneer into making paper from Australian eucalypts, was General Superintendent of Australian Newsprint Mills Limited from 1938 to 1956.

By 1922 they were at a stage where they needed to pilot their methods on a manufacturing scale. They had a machine built at Fremantle and shipped to the Australian Paper Manufacturers Limited plant near Geelong, where paper was then being made from overseas pulp. For the first time, Benjamin actually saw a real paper mill.

After the Geelong tests, Benjamin turned his attention to newsprint. At that time Australia was using about 70 000 tonnes of newsprint a year, all of it imported. Such a large quantity of a single grade seemed to offer economic opportunities worth pursuing.

Benjamin's early investigations into making newsprint from eucalypts concentrated on finding the right balance between mechanical pulp, wood that has simply been ground up and bleached, and chemical pulp, wood which has been chipped and cooked with chemicals.

Because Australian-made newsprint would probably require a larger proportion of chemical pulp than overseas-made newsprint, Benjamin and his team concentrated on finding a cheaper way of making it. They turned from the caustic soda to the sulphite process, which gave such encouraging results that in the spring of 1924 they shipped 200 tonnes of white stringybark and myrtle logs to England for large-scale tests.

The only mill in England then using the sulphite process refused to pulp the logs, so they had them shipped to Holland, where they had found a more co-operative mill. On 17 March 1925 they made the first production run of newsprint from Australian timber.

Back in Melbourne with their precious newsprint they had a few rolls tested on the presses of the Melbourne *Herald*. 'A more unpromising material for use on fast rotary presses had never been tried before or since,' Benjamin wrote.

But his financial backers, Amalgamated Zinc (De Bavays) Ltd, decided to go ahead with manufacture of newsprint in Australia from eucalypt timber if they could get suitable tariff protection. Naturally enough, the newspaper proprietors opposed the application when it came before the Tariff Board. The Board granted a bounty of £4 a ton, but the company said the project would not be viable with anything less than £6 a ton.

Benjamin and his team set out to see if he could close the gap between the

Three-pocket grinder used at the Kermandie pilot mill in the late 1920s. Wood was ground to pulp in these.

A production refiner with 36-inch surfaces.

£4 offering and the £6 needed. At least they had overcome the quality problem, for when they made a sample book of their own laboratory and imported newsprints and took them to the *Herald* nobody, from the press-room superintendent to the general manager, could tell which was which.

Amalgamated Zinc in 1927 built a pilot pulp and paper mill at Kermandie on the Huon River in southern Tasmania. It enabled Benjamin to work out the technicalities and economics of making paper from Australian timber, but the plant was closed when the depression hit in 1929.

The whole picture changed in 1932, when the Melbourne *Herald* decided that the best defence against a company making high-priced newsprint behind a tariff wall was to start such a company itself. It formed the Derwent Valley Paper Company, got forest concessions from the Tasmanian Government, and enticed Benjamin to join it. Its first production run was in February 1941. Today the company is Australian Newsprint Mills Limited.

In the meantime, a group of Melbourne mining companies had joined together to finance a new paper mill at Burnie, Tasmania, which used local timber as well as imported pulp to make a range of printing and writing papers.

28

A vehicle for taking the family to church on Sundays, the pigs to town on Mondays

A FARMER WHO wrote to Ford Australia in Geelong in 1933 started the chain of events that led to design and production of the world's first utility.

Unable to afford both a car and a truck, the farmer described a new type of vehicle. 'The front is the coupe, to suit my needs for taking the family to church on Sunday,' he wrote. 'The back is to be the roadster-utility box so I can take my pigs to town on Mondays.'

Ford's Managing Director asked the production superintendent who asked the chief body engineer, Lewis Bandt, 'Can it be done?' Bandt felt it was one of those requests that would soon fade away, but he was surprised to find, when he put it on paper, that a surprisingly well-balanced vehicle came out of the idea.

Once Ford management saw the drawings, they decided they wanted to see some real vehicles. Months of work went into making the first prototypes by hand from a mixture of production and hand-made parts. When the sales people saw them they said 'We could sell hundreds of those tomorrow!'

So started the coupe-utility, 'coupe' for the front two-seater part, 'utility' for the hard working rear-section. This kind of vehicle was soon simply known as the 'utility' or 'ute'.

1934 Ford utility, the world's first production-line ute, designed and made in Geelong.

A Ford utility of the 1960s.

There had been small trucks before, of course, and even some hand-made models which took the front half of a canvas-topped tourer and added a truck-type body. Few of these had any claim to style or comfort. But Ford Australia turned out the first production model that had the style and comfort of the sedan and the utility of the buckboard back.

The first utilities, like the sedans of the time, had wooden frames. But by the late 1930s both utilities and sedans were being made with all steel bodies.

The utility idea spread rapidly from Ford to other makers and from Australia to many other parts of the world. Whenever a new model sedan appeared, a matching utility was usually close behind.

29

Army brass resisted production of the Owen gun, but it became 'the Digger's darling'

WHEN WOLLONGONG INVENTOR Evelyn Owen's sub-machine gun finally went into production it was such a success that Australian troops in New Guinea called it 'The Digger's darling'. Yet for two years the army seemed to be doing everything possible to prevent production of the Owen gun.

Born in Wollongong in 1915, 'Evo' Owen showed an interest in guns from a very early age. He was constantly experimenting with homemade firearms of one kind or another, a habit which twice involved him in near fatal explosions. On one occasion one of his 'inventions' spattered his abdomen with shrapnel when it exploded, leaving him scarred for the rest of his life. On another occasion a bullet entered the front of his abdomen and came out the back.

By 1939 he had developed a sub-machine gun, and showed it to an ordnance officer at Victoria Barracks in Sydney. He was told that the army was not greatly interested. Sub-machines were more for gangsters than soldiers, more American than British or Australian.

This attitude to sub-machine guns changed quite rapidly when World War II started, for the German army used them widely in their successful drives across Europe. But the Australian army still said it was not interested in the Owen gun. It preferred to wait for the Sten gun, a sub-machine gun that was being developed in Britain.

On the outbreak of war, Evelyn Owen joined the Australian Army. While he was in training, an Asian firm offered to buy the plans of his gun for the Japanese army, which fully realized its likely value in jungle fighting. Owen refused to sell.

In September 1940, when he was on final leave before going to the Middle East, Owen put the working model of his gun in a sugar bag, where his neighbour, V. A. Wardell, who happened to be the manager of the Lysaghts works at Port Kembla, found it.

Wardell soon realized just how rapidly the gun could be produced in large numbers, and brought it to the attention of Essington Lewis, the BHP chief who was also in charge of Australia's munitions production. He sent Owen to see Captain Dyer, Secretary to the Army Central Inventions Board in Melbourne.

Ironically, the Board was not working too well because of the war. Dyer had trouble persuading his superiors to see the Owen gun's potential. He went outside normal channels, however, and had some test models made at Lysaghts, and these worked well.

After questions were asked in Parliament about the gun, the Government took the subject out of the hands of the army and ordered 100 Owen guns from Lysaghts, much to the chagrin of the Military Board.

The Owen gun was tested against the Sten and Thompson guns at Long Bay Range on 29 September 1941 and it performed very well. Under some conditions, such as after being immersed in mud and water, it performed better than either of the other two.

The Owen gun was very simple in its design and construction, which meant that it worked well under difficult service conditions. It could be taken apart and put together again very quickly. Its system of fire control prevented waste of ammunition.

Its high muzzle velocity helped to make it a very effective short range weapon. Its magazine was on top, which meant that the gun's centre of gravity was on its longitudinal axis, a fact that gave it greater accuracy when it was set to fire automatically. This placement of the magazine also meant that if mud got into the

A cutaway diagram of the Owen gun.

Evelyn Owen, right, discusses features of his gun with D. West, a sight setter and gun tester.

gun around the breech it would drop straight to the ground. If it got into the magazine it either fell out or was pushed out by the magazine spring. The Owen gun's ability to operate under muddy conditions made it superior to other submachine guns in the jungle.

Eventually Lysaght's got the Owen gun into production and made 45 477 of them at their Port Kembla plant, even though the army seemed determined to put every obstacle it possibly could in the way.

Three submachine-guns. Owen (9 mm) top, Thompson (.45 in) centre, Austen (9 mm) bottom.

The poor handling of the Owen gun by the Army Central Inventions Board was one of the main reasons for the setting up of the Army Inventions Directorate, a civilian body that was more realistic about inventions and more helpful to inventors.

In December 1943, when the Ordnance Board of Britain tested the Owen in competition with five other guns, the Owen was rated first in four of the five tests and first in over-all order of merit.

War Correspondent Damien Parer reported that the Owen gun was the greatest morale builder the army had, and a wartime poet wrote:

> Sweat and tears! said Churchill,
> Tanks, tanks for de Gaulle,
> Planes for Chiang Kai-Shek,
> Who wants them big or small.
> 'Struth,' said the private,
> In his true Australian drawl,
> 'All I want's a little Owen,
> It's the daddy of them all.'

The Owen gun saved Australia £2 000 000, according to some estimates, and almost certainly the lives of thousands of her troops. Yet Owen did not benefit markedly from it. The government paid him five shillings a gun royalty, which amounted to about £11 000. The high wartime taxes are said to have taken £9 000 of this, leaving Owen with £2 000, not enough to cover his expenses.

He certainly lived like a poor man after the war, retiring to a family timber property, where he worked the timber mill by day and lived in a one-room hut, with only a radio for company. He died in 1949 of a combination of a gastric ulcer and a heart condition, plus pneumonia.

The Owen gun survived long enough to be used in the Korean War in the 1950s.

Wollongong Ex-Services Club recalls Owen's memory with a mounted replica of his gun and a plaque. Owen's sister, Eleanor O'Donnell, recalls that her brother was 'a very interesting and charming young man with a gentle nature and many friends.' He was exposed to an enormous amount of publicity 'but remained himself, and totally without conceit'. He received many letters from soldiers about his gun.

30

Melbourne pieman asked Uncle Sam for £5 000 000 royalties on his exploding bullet

JOHN POMEROY was one inventor who thought big, at least when it came to royalties. At the end of World War II he demanded £5 000 000 from the United States for his idea of an exploding bullet.

Pomeroy had produced his first exploding bullet over 40 years earlier, in 1902, in his native New Zealand. When New Zealand army experts tested it in 1904, they were very enthusiastic about its potential.

Pomeroy demonstrated the bullet with an old Martini Henry rifle. In one demonstration he stretched a sheet of paper a metre behind a tree. When he fired a bullet into the tree, fifty of its exploding pellets passed through the paper. This bullet was powerful enough to kill ten men with one shot.

Its secret was centrifugal force. As it spun it threw the chemical elements in its lead case so violently apart that they exploded on the lightest contact. A bullet that touched a man lightly on the arm would kill him as surely as one that went straight to his heart.

In spite of the New Zealand army experts' enthusiasm for the exploding bullet, nothing was done about it. Pomeroy even went to England to see if he could interest the British army but could not get any real response there either.

Only when World War I made the need urgent did officials act. Even then they did not hurry themselves.

With Zeppelins raining bombs on England, Pomeroy pestered the War Office to try his bullet against them. He was told there was no need, as the war would be over within a few months. But soon after he got back to Australia, where he now lived, Pomeroy had an urgent call from the War Office asking him to go back to England. He was supposed to sail from America to England on the *Lusitania*, which was sunk off the coast of England by a German U-boat, but his train was delayed and he missed the boat.

In England he waited another 16 months before anything happened. Then on the first night his bullet was used against the Zeppelins it brought one down at Potters Bar. The skies over England were soon clear of the raiders, and the

'Pop' Pomeroy with the horse that pulled his Melbourne pie cart.

British Government paid Pomeroy £25 000 in royalties for his bullet. He declined a knighthood, but his wife was made a Dame of the British Empire.

Pomeroy was a fairly prolific inventor throughout his life. The money he made on his successful inventions all disappeared into his efforts to promote the unsuccessful ones.

His first invention was a non-slip clothes-prop, invented about 1884 in Invercargill, New Zealand, where he was born. He came home from school one afternoon to find his mother in tears because her clothes-prop had slipped and allowed her washing to trail in the mud. He fixed a clip of fencing wire to the top of the prop to attach it firmly to the clothes-line. Then he fitted two prongs to the bottom to stab into the ground. Pomeroy's father patented the idea for him and sold the rights to a local hardware merchant for £50.

After a stint as an apprentice engineer and seeing the world as a seaman on a windjammer, Pomeroy settled down in New Zealand to some steady inventing.

He devised a fountain pen that filled itself when it was dipped in ink. It was efficient and practical, but before he could get it into production a similar product came on the market. Pomeroy said the other inventor had seen a model of his pen, but there was nothing he could do about it.

About that time the fast bowlers in a touring English cricket team had been unnerving Australian and New Zealand batsmen. Pomeroy invented a pneumatic cricket pad that took the terror out of even the fastest delivery. He tried to sell the idea to a Melbourne sporting goods firm who agreed to try it in a match.

The pneumatic pad did everything it was supposed to do and more. Every fast ball that hit it bounced off and went straight to the boundary for four leg byes. One batsman using the pad accounted for 450 of these, and it was soon abandoned.

All that was before Pomeroy invented the exploding bullet.

After World War I Pomeroy came home via the United States. He spent four months there selling his 'elixir of life', a patent medicine he had developed. It was supposed to keep people fit after 40. He cleared $50 000 on this venture.

Pomeroy settled in Melbourne where he invented a fire bar for stoking furnaces, a blue flame cooker, a non-burring sharpener for razor blades, a gadget for cutting and combing hair at the same time, a regulator for shearing, headlamp dimmers, and a rabbit trap that did not hurt the rabbit.

He also worked out a method for taking the bitterness out of oranges and for freezing oranges in great blocks. The idea never reached the production stage because Pomeroy held out for a higher royalty than commercial interests were willing to pay.

Pomeroy managed several hotels in Melbourne in the 1920s, but they did not earn enough to finance several of his inventions that failed, including one to cure cancer. He decided to operate a pie stall in Melbourne, and after furious battles with municipal authorities managed to get it into operation at a permanent site on railway land. 'Pop's Pie Car' became a Melbourne landmark.

In 1937 Pomeroy's explosive bullet came into the news again. Chinese leader Chiang Kai-Shek invited him to China, then at war with Japan. Pomeroy sold the Chinese a new and improved version of his explosive bullet.

During World War II the United States army used explosive bullets, including the 'blue nose' 0·5 calibre shell. It contained a small armour-piercing cylinder which continued in flight after the main shell had hit and exploded. The 'blue nose' shell was an important factor in the Allied victory in the Coral Sea, which saved Australia from Japanese invasion.

The United States rejected Pomeroy's claim for £5 000 000 in royalties for the explosive bullet. It said they had been working on it long before Pomeroy sold his bullet to the Chinese.

Pomeroy died in 1950 of cancer. He had spent his last few years on ceaseless but fruitless work in search of a cancer cure.

31

From bullet-proof radiators to plywood rafts

AUSTRALIAN INVENTORS worked overtime during World War II. Although none of them became as well known as Evelyn Owen and his sub-machine gun, they turned out a steady flow of such useful inventions as bullet-proof radiators, mine detectors, plywood rafts and signalling mirrors.

A bullet in a radiator could put a whole vehicle out of action. The established method of bullet-proofing was armoured shutters which could be closed when necessary. These usually kept bullets out, but they blocked the airflow and prevented the radiator from working as it was meant to.

The Army Inventions Directorate therefore arranged to develop the idea of P. S. Barma for a bullet-proof radiator. Barma proposed fitting a series of vertical armour plates edgeways in front of each of the water-carrying cores in a radiator. These would deflect bullets away from the vital cores, which held the water. Bullets might pass through the heat dissipation fins, but this did not seriously impair operation of the radiator.

Tested in May 1943, a radiator fitted with these fins was not damaged by rifle and machine-gun fire at 35 metres. Development work however was slow, and by the time the radiator was perfected so many vehicles had been fitted with conventional radiators that the army decided not to introduce the new one.

Rather more successful was Miller's mine detector. Miller saw a film of Australian soldiers searching for land mines in the Western Desert by prodding the ground with bayonets. He designed a mine detector with a tuning coil which formed a searching loop attached to a light rod. This coil was coupled to one wireless valve that acted as an oscillator. Any change in the loop capacity inductance (such as caused by the presence of a mine or other metal object) was detected by another valve, which acted as a regeneration receiver.

Miller's invention became known to the army as 'Detector Mine No. 1 Aust.'. It gave much valuable service before it was replaced with something better.

Early in the war, Corporal H. W. Jeffrey of the Royal Australian Air Force proposed a robot tankette, a small tracked vehicle about a metre long, half a metre wide and half a metre high, and weighing about 200 kg. The tankette, designed to carry explosives, had small battery-powered motors and trailing cables for steering. It could be sent to detonate mines or to blow up bridges. It seemed to work, but the army knew that similar British and German tankettes had not been a great success and did not persist with the idea.

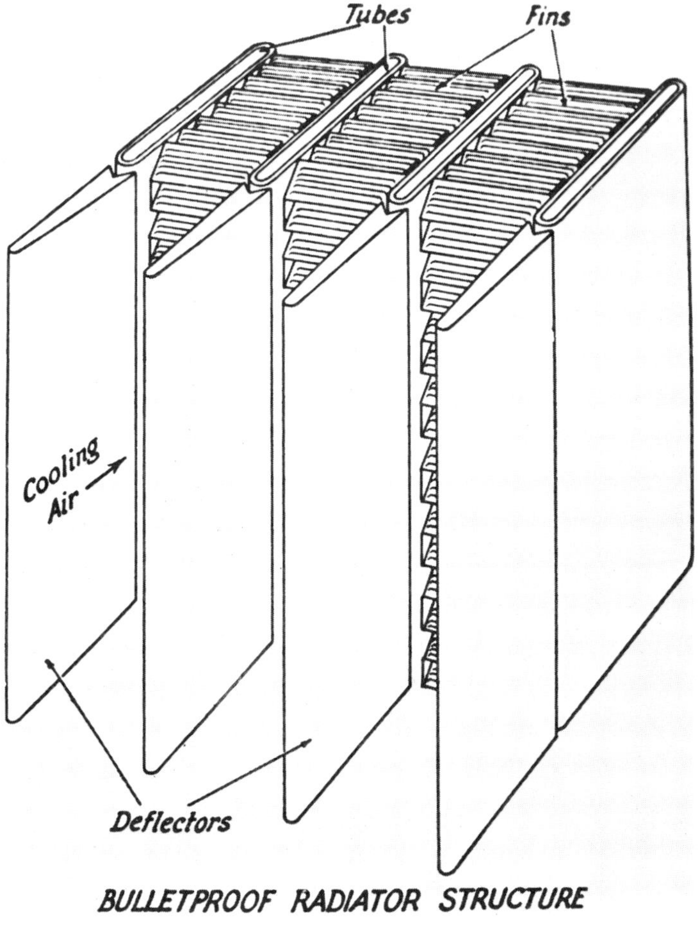

Australian design for bullet proof radiator for motor vehicles.

One useful wartime Australian invention was the cellular plywood float raft for ships. When supplies of the highly buoyant South American balsa wood were interrupted by the war, Sydney furniture manufacturer F. Mills designed a raft of about 26 cells of water resistant plywood. Each raft could support two men, even if half the cells were destroyed. Over 2 000 such rafts were made.

Another life-saving Australian invention was Tom Robertson's signalling mirror. This used the well-established heliograph principle, which Robertson had seen warships using off Gallipoli. When he read that some survivors of a torpedoed freighter had signalled a searching aircraft with a tobacco tin, he recalled these ships, and invented a mirror that enabled its operator to direct a flash of light directly at a searching aircraft by lining it up in the sight. About 20 000 of these mirrors were made and distributed. After the war the mirrors were used in outback Australia to attract Flying Doctor pilots to the best landing spots.

The inventions described here were only the tip of the iceberg. During World War II so many people sent in ideas they thought would help the war effort that the War Cabinet set up a Central Inventions Board and later an Army Inventions Directorate to winnow the chaff of thousands of useless and fantastic ideas from the few grains of worthwhile inventions.

All told the Directorate examined 21 645 submissions, of which it referred 3 686 for assessment and expert advice. Of these, 175 devices were actually built and tested, and 127 finally accepted for development. This was only 0·58 per cent of the total submitted, but even this small yield compared favourably with those in other countries.

32

George Shepherd's love of bridge saved the world a lot of pushing and shoving

---

George shepherd loved bridge. But he felt that much of the enjoyment in the game was cancelled by the uncomfortable furniture most people used when they played cards.

So in the 1930s the Melbourne oil company engineer decided to do something about it and before he had completed his project he had given the world the Shepherd castor, which meant people could roll their furniture about without all the pushing and shoving that had been involved with the unsatisfactory old style castor.

George Shepherd began by fitting his own bridge room with a new adjustable steel table and four comfortable armchairs, all of which he designed himself. These were a big improvement, but Shepherd soon saw that their main problem was the old style five centimetre castors fitted to them. He made some 10 centimetre castors of similar design. They rolled more easily than the smaller ones, but they did not align themselves as well.

Size was not the real problem, Shepherd decided. The design for castors had not been changed for 200 years and they had several faults which he decided to overcome. Having the pivot above the castor provided too much side leverage, so he tried to overcome this by fitting two wheels, with the pivot bearing between the wheels in front of the axle. This system worked well on a good floor, but on a slightly irregular floor all the weight finished up on one wheel, which prevented it from castoring.

Shepherd's next idea was an inclined single wheel, which meant the pivot could be housed inside the wheel. He made a series of such wheels, testing diameters, offsets, treads, etc., all of which gave better results than any previous castors.

These 'flying saucer' castors, as Shepherd called them, soon had a severe test. When Shepherd's wife broke her ankle, he fitted them to a chair so she could push herself around the house with a walking stick. The new castors took her smoothly over carpets, linoleum, mat edges and concrete for weeks.

Convinced that he had discovered something worthwhile, Shepherd now

George Shepherd.
Uncomfortable bridge chairs started him on the road to inventing the Shepherd castor.

worked on the problems of lubrication, dust-proofing and general appearance. As a result he produced the first Shepherd castor to look like the version that finally came on the market. It was a 10 centimetre castor with an inclined domed wheel giving room for a long bearing with an oil trap, and a domed cover fitting inside the wheel to enclose the working parts.

By now Shepherd could see that his idea had potential far beyond his bridge room furniture. To prove the worth of his new design he hand-made 60 castors and fitted them to furniture throughout his house. They were still going strong 20 years later, in 1959, when Shepherd replaced them with a later, better-looking version.

World War II slowed down Shepherd's development of his castor, but he steadily refined it in readiness for mass production. He reduced the diameter gradually to find the smallest size that could carry a 40 kilogram load. He fitted a set of seven and a half centimetre castors to a trailer carrying over 300 kilograms and towed it for several kilometres. He left four castors standing for six months with a static load of one tonne. They showed no ill-effects.

At the end of the war Shepherd licensed an old friend, Mark Cowen, to make and sell the castor. Market acceptance was slow, but people gradually came to appreciate the advantages of the Shepherd castor and insist on it when they bought furniture.

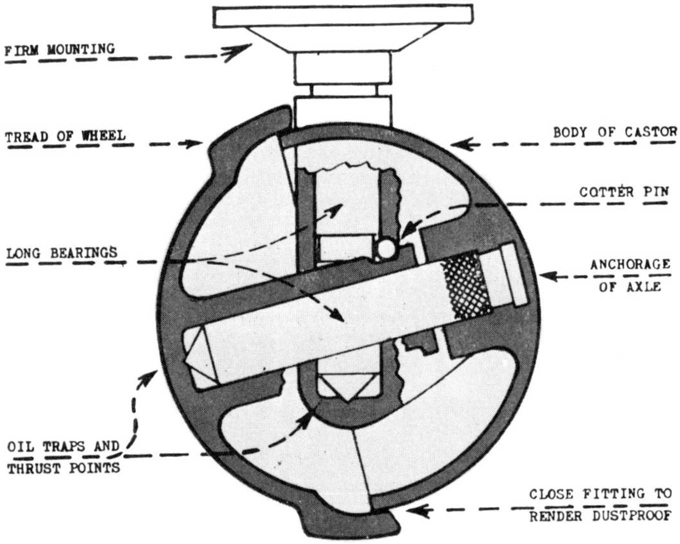

Sectional view of the Shepherd castor.

Distinctive features:

1. Free mobility in any direction
2. Firm mounting without wobble
3. Lubrication gives silent smooth action
4. Clean dust-proof form of great strength
5. Good finish and long life

Shepherd continued to develop his design. The dust cover became an integral part of the castor, and allowed sensitive castoring on a six and a quarter centimetre model. The number of parts was reduced from eight to three plus a cotter pin, which gave a stronger and more durable castor at a lower price. The new design came on the market in 1948 and gradually replaced the earlier model. Later developments included a castor with a rubber tread, which is better for floors without carpet, a castor with a brake, and a junior castor, a slightly smaller version for lightweight furniture.

From Australia the Shepherd castor rolled its way around the world.

33

Lance Hill put the hoist in rotary clothes lines

SOON AFTER LANCE HILL was demobbed from the army in 1945, his wife asked him if he could think of something better than the old clothes line and prop she had in her Adelaide back yard.

Hill's thoughts turned to the rotary principle which had been tried before in Australia but had never caught on. To replace the traditional prop, he decided on a winding mechanism to lift the clothes to catch the available breezes. Being a motor mechanic, he naturally enough thought of a crown wheel and pinion and that is what he used.

Neighbours, impressed with the obvious convenience of the 'windmill', as it was then called, asked Hill to make them one, and orders started to flow in. Harold Ling, Hill's brother-in-law, suggested that with sound marketing there was a bright future for the rotary clothes line. Early in 1946, with a little capital provided by Lance Hill's father, a partnership was formed.

Papa Hill, who died in 1969 at the age of 94, delivered the clothes lines in a wheelbarrow every Tuesday. By the end of 1946, orders were coming in at such a

Lance Hill (centre) with his brother-in-law, Harold Ling (left), who later became the chairman and managing director of the public company formed to make Hills Rotary Clothes Hoists, and Jack Short, who became a director of the company.

Wind up gear of an early model Hills Rotary Clothes Hoist, before assembly.

rate that investment in a more sophisticated vehicle seemed to be warranted. The firm bought some clapped out army trucks.

'Mass production' started in 1948, after purchase of the local blacksmith's shop at Glen Osmond. Postwar shortages of materials and the high cost of water pipe, which was used for the main standard and arms of the rotary driers, led to acquisition of a firm that made metal tubes. It was this tube mill which had so much influence on the future development of the company.

Steady growth of the business led to steady improvement of the Hill's Rotary Clothes Hoist, and a series of patents were taken out.

The Hill's Rotary Clothes Hoist was well received by the Australian public right from the start. All its teething problems were related to shortages of capital, raw material and plant rather than to any shortage of customers.

First incorporated as a private company in 1948, the firm went public in 1955. Lance Hill retired for health reasons just before the public float. Harold Ling retired in 1965, retaining the role of chairman until his death the following year, when it was taken over by his son.

34

Metcalfe and Wright put some tone in xerography

In the 1950s two scientists at the Defence Standards Laboratory in Adelaide, Ken Metcalfe and Bob Wright, produced some notable improvements in the xerographic process, first invented by Chester Carlson in the United States in 1937.

Carlson's original process was limited to line reproductions, those in which the image was a mixture of black and white, with no grey areas such as required for the reproduction of photographs. It was not suitable for half-tone reproduction, in which pictures are made up by a series of dots, as in newspapers. It was even less suitable for continuous tone reproduction, such as that of photographic prints.

The Defence Standards Laboratories first became interested in xerography when it seemed the process might be adaptable to industrial and medical radiography, for example in X-raying stored ammunition. To make this possible, however, the process would need to produce continuous tones.

Metcalfe and Wright in 1952 worked out a system of liquid rather than dry developers. These contained pigment particles that were much finer than those used in the dry process. In conjunction with a printing plate that had varying levels of charge, the liquid developers could provide continuous tones.

From the outset it was clear that liquid developers brought to xerography a much greater range of applications. Liquid developers could be produced in any colour, so images could be overprinted to give multicolour reproductions.

Since their introduction in 1952 the liquid developer processes have been under continuous study and many special types of liquid developer have been devised for special purposes. By 1967 about 30 patents had been taken out for major innovations in xerography made by the Defence Standards Laboratory. Licences have been granted to Australian, American, European and Japanese firms to use them.

One Australian licensee has developed a commercial system of colour-proofing screened half-tone colour separations. In colour printing, four different plates are used, one providing the reddish hues, one the blue, one the yellow, and one the black. This involves an intermediate stage of making colour separations

An aerial mapping photograph printed as a modern electrograph.

onto four sheets of film, from which the plates are made. Until the new system was developed the colour separations could be checked only after the plates or blocks had been made. The new system, based on the Defence Standards Laboratory patents, enabled the separations to be checked and adjusted very quickly and cheaply before the plates or blocks are made.

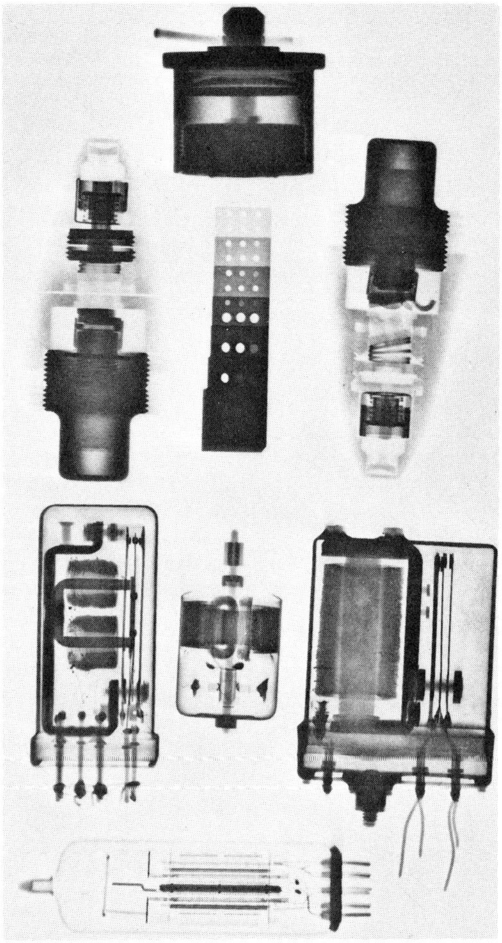

An electroradiograph of industrial components. Without tone this would be meaningless.

An American defence contractor uses the Australian process in his mobile five-colour electrophotographic map-printers for field service. Each machine can produce 2 500 five-coloured maps per hour from easily stored 70 mm colour transparencies. This greatly reduces the problems of supplying maps to combat areas.

35

Ken Gaunt finally took the backbreak out of cane-cutting

UNTIL THE LATE 1950s, Queensland's enormous sugar cane crop was cut almost entirely by hand. Gangs of men moved from farm to farm, stooping low as they wielded savage cane knives to sever the cane stalks at ground level. They cut the top off the stalks, laid the stalks out flat on the ground, then picked them up in bundles and loaded them on small rail waggons for transport to the mills.

In the fierce tropical heat their copious sweat mixed with the soil of the cane-fields and the ash of the foliage, usually burnt off just before each field was cut.

Some farmers did have a mechanical loader, a sort of bucketless front-end loader with a grab. With this the farmer could pick up the cane and load it into trucks. There were even some mechanical cutters of a kind. But these were mostly designed to copy the manual operation, cutting off the cane and laying it on the ground. They worked well enough in cane that was erect, but not in cane that was fallen and twisted, as it often was.

In 1955, Canadian Tom Ritchie, newly appointed managing director of Massey-Ferguson (Australia) Limited, saw manual cutters at work in the Queensland canefields. 'Those cane-cutters are working in hell,' he said. 'There just has to be a way to eliminate that torture.'

Ritchie sent his superintendent of engineering design and research, English born Ken Gaunt, to Queensland to study the possibilities of mechanical harvesting.

Gaunt soon had a list of basic needs for a commercially successful mechanical sugar harvester. It had to be capable of harvesting crops in every condition, and not just straight standing cane. It had to be a true harvester, delivering the cut cane loaded ready for transport to the mill, rather than just dumping it on the ground. It had to be produced at a cost the average small Queensland cane farmer could afford, and it should preferably be tractor mounted in a way that left the tractor available for other work. On the other hand, it needed to be independently powered, so that it did not depend upon a regular forward speed. If it came upon a particularly heavy patch of cane it needed to be able to slow down to handle it.

At this stage Ken Gaunt apparently still contemplated a machine that harvested each stick of cane in one piece in the traditional way. While still pondering the weighty problem of how to deliver short or long, straight or bent, brittle or tough cane sticks with constant ease into a carrier behind the cutting mechanism of an embryo machine, Gaunt one day saw the hand cutting of a crop at McKay. It was a revelation to him.

The field had been flooded twice. The cane had gone down flat both times,

Ken Gaunt.

then grown upright again before a high wind finally flattened it a third time. Each stick was about five metres long and had two right-angle bends.

The cane cutters chopped each stick singly near the ground then pulled it from the other end to remove it from the tangled mass. Often the stick would not come free and had to be hacked loose at the other end. The cutters threw the sticks into an untidy mass and lifted them to the trucks with a front-end loader.

Gaunt made two decisions from this experience. First, no machine could ever

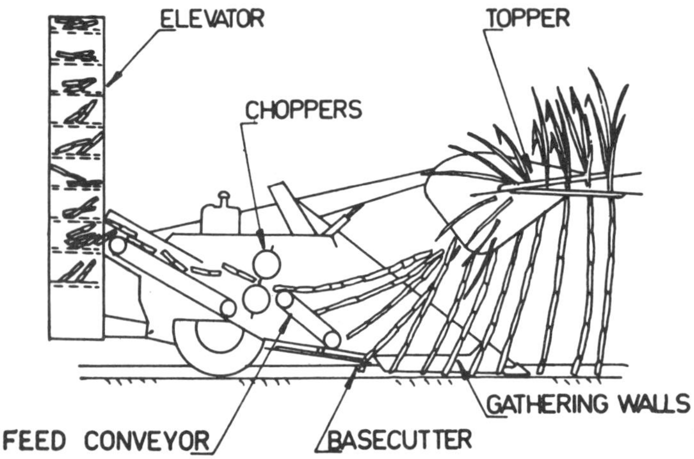

Simplified diagram of how a mechanical sugar cane harvester works.

handle a crop like this as whole stick cane. Second, the bottom end of the stalk was the business end. If it could be picked up and pulled into a machine the rest of the stalk would follow it in.

Because of the great variety in size and condition of cane, any machine would need to reduce it to uniform length pieces for elevation to some sort of bin. If chopping was feasible, and chopped cane was acceptable to the mills, then the variety, weight and condition of the cane did not matter.

Gaunt and his team worked out a design they thought would cut and chop

An early Gaunt chopper harvester.

cane and in October 1956 they tested it at the property of Cyril Jackson, near Mackay.

'We didn't cut much cane at that exercise,' Gaunt said later, 'but we found out why the machine wasn't cutting it, which was just as valuable.'

In 1958, Massey-Ferguson put its prototype cane-harvester through three days of public demonstrations. General opinion among growers was that it was very close to being a practical means of mechanical harvesting.

In the 1959 season Massey-Ferguson put six harvesters in the field at various places in Queensland. This gave the company operating experience in widely varying areas and conditions, including irrigation areas at Ayr, stony hillsides outside Mackay and excessively wet blocks at Mulgrave.

After the season they reconditioned these six machines and built another six. They shipped one to the United States, leaving eleven in operation in Queensland in 1960.

By 1961 the Massey-Ferguson 515 model cane-harvester was coming off the production line at the Company's works at Sunshine, Victoria. In that year about 235 000 tonnes, equal to 2-3 per cent of the Australian cane crop, was mechanically harvested by Massey-Ferguson chopper harvesters. Today Australia's sugar crop is cut almost entirely by mechanical harvesters.

36

Mervyn Victor Richardson made a million from his mower

ONE DAY IN 1951 Sydney university student Garry Richardson decided to make some money by cutting lawns during his vacation.

As he set out to start his new business, he asked his father, 57-year-old engineering firm salesman Mervyn Victor Richardson, to help him load the family mower into a motor-cycle sidecar. It was a power mower, just the thing for somebody going into the lawnmowing business. But it was one of the heavy, cumbersome, cylinder types, the usual kind of powered mower at the time.

Not that there were too many powered mowers at all. Most Australian men lived in dread of the weekly chore of pushing a hand mower around their suburban yards.

Mervyn Richardson began to think about his son's idea of making money from lawnmowers. But the father's ideas were working on a grander scale. Why, he asked himself, did a powered mower have to be so big that the average householder could not lift it, and so expensive he could not afford it?

He thought of the light, inexpensive vacuum cleaners which most Australian women were then pushing round indoors, and wondered why Australian men were not pushing round something equally light and inexpensive outdoors.

True, there were electric mowers. But they did not seem to be the answer. There was always the risk of running over the power lead and cutting it, and most electric mowers of the time were far from rugged enough to handle the tough conditions encountered in many an Australian backyard. Richardson wanted something with go-anywhere ability that was also light and inexpensive.

He decided that a small petrol engine would probably be the ideal power unit, and started sketching what it would be like. His job as an engineering supplies salesman had given him some idea about engineering, and he was a reasonable sort of handyman. Even though he was 57, he had just completed a draftsman's course, getting through the usual three years' work in five months. He had once helped his brother build an aeroplane, had made jewellery and had as a boy done signwriting work.

Richardson's first mower was a cylinder type, just like the few others on the

Mervyn Victor Richardson. Profits from his lawnmower enabled him to indulge his interest in flying.

market. But it was light and inexpensive, selling for only £67-10-0. He called it the Victa, after his middle name, Victor, and it was the Victa 14 because it had a 14-inch cut. Richardson put the mowers together at his home in the Sydney suburb of Concord, relied mainly on word of mouth for advertising, and gave potential customers demonstrations on his own lawn.

Although the mowers sold as fast as Richardson could make them, he was far from satisfied. They lacked the power to attack a good growth of paspalum successfully. What he really needed, Richardson decided, was a mower with a rotary action. He set about to design and build a prototype.

Finally he had something worked out. He shaped and bolted and welded metal into a frame on wheels, mounted an engine on top, attached some blades under it, and fed the whole thing with petrol from a peach can.

Victa 500 twin lawnmower, with grass-catcher.

When it was finally finished, the aspiring inventor allowed his wife the honour of being the first to use it. She pushed it like a child's stroller into some long grass beside the fence, and the grass seemed to melt before the onslaught.

Richardson's new mower not only sliced its way through tall grass. It could shave a good lawn as nicely as one of the cylinder type mowers, and it could trim right up to the edges of garden beds and within a whisker of trees and shrubs.

The first Victa rotary lawnmowers were assembled in the summer of 1952-3. As the Richardson family assembled the mowers in their garage people queued up outside waiting to buy them. As each mower came off the end of the 'assembly line' a customer stepped forward, Richardson demonstrated the mower on some of his lawn, and sold it for cash on the spot. One of his major problems was that he kept running out of lawn that needed cutting.

Richardson gave up his job so he could devote himself full-time to his lawnmower business. In February 1953 he moved to a shed in Mortlake, and later to premises on Parramatta Road.

In August 1955 the lawnmower enterprise, now a family company, moved to a new factory at Silverwater; by then it had 100 people working for it. A week after the move a disastrous fire raged through the factory, destroying £30 000 worth of machinery.

The fire occurred right at the beginning of the grass-growing season, the best time to sell mowers. The firm did everything possible to get back into operation, and within a week had partial production restored. Some parts even came by air from England. Within a month the plant was back to full production. The major line at this stage was the Victa 18-inch Rotomo (rotary mower).

By the end of the firm's fifth year, it was turning out 3 000 Victa mowers a week, one coming off the production line every 48 seconds at maximum production.

Within a few more years it was no secret that Mervyn Victor Richardson was a millionaire. And that, in the days when the classification meant having a million pounds, not a million mere dollars. He was now able to relax at such diverse hobbies as ballroom dancing, speedboat racing, flying and table tennis.

Richardson spent £130 000 on building a house on Pittwater at Sydney's Palm Beach. It had circular rooms, sunken baths, and a swimming pool that extended into the lounge. There was also a hangar-boathouse to house a four-seater amphibious plane and a speedboat. Richardson sold his dream house in 1966. At 52 squares, he said, it was a little too big.

37

Alan Walsh speeded up chemical analysis with the atomic absorption spectrophotometer

SCHOOLCHILDREN WHO persist with chemistry long enough come to learn that each chemical element gives a flame a characteristic colour. One primitive analytical technique is to dip a wire loop in the sample, hold it in a bunsen burner flame and watch the colour of the flame.

Lancashire-born Dr Alan Walsh put a similar principle to work in the atomic absorption spectrophotometer, a high speed instrument that has solved analysis problems in medicine, mining, pollution control and a host of other applications all over the world.

The story of the atomic absorption spectrophotometer began in the early 1950s. Then as now Dr Walsh was with the Chemical Physics Division of the CSIRO. His techniques of analysis were rather more advanced than a loop of wire and a bunsen burner and what his eyes told him, but the commercial infra-red spectrophotometer he was using depended on the same principle: it depended on the light emitted by a vapourized sample.

Atomic emission spectroscopy had its limitations. It could detect only seven or eight elements, because even in the hottest laboratory flames only a few atoms became excited enough to emit light. Even then there are many practical problems in analysing and measuring the light they emit.

Walsh was pondering the problems of atomic emission spectroscopy one Sunday morning in March 1952 as he dug in his garden in the Melbourne suburb of Brighton. Suddenly he dashed to his phone to call his colleague John Shelton.

'We've been measuring the wrong bloody thing!' he said. 'We should be measuring absorption, not emission!'

Although only some elements can be excited enough to emit light, all atoms in their normal state absorb light, and this can be measured accurately.

Early next morning Alan Walsh set up a simple experiment. By morning tea-time he had a successful result, the basis of what has been called the most significan advance in chemical analysis this century.

Walsh and his colleagues gradually overcame the technical obstacles one by one and by 1954 had built a prototype atomic absorption spectrophotometer.

They gave a public working demonstration in March 1954 at Melbourne University as part of an exhibition of scientific instruments.

Apart from the technical obstacles, Walsh also had widespread indifference to overcome. Many people saw atomic absorption spectroscopy as a scientific curiosity rather than a practical analytical method.

'If this technique is as useful as you say it is, why isn't it being used right here in the United States?' the head of an American scientific instrument corporation asked. Walsh convinced him that the United States could learn a lot from Australia on atomic absorption spectrophotometers, and the corporation has become one of the world's largest makers of atomic absorption equipment.

Walsh originally used a bunsen-type flame to produce atomic vapours from the sample. To measure the concentration of a particular element, say sodium, he put on one side of the sample a lamp that gave out light at a wavelength characteristic of sodium. On the opposite side he put a photocell which analysed

Alan Walsh with an atomic absorption spectrophotometer.

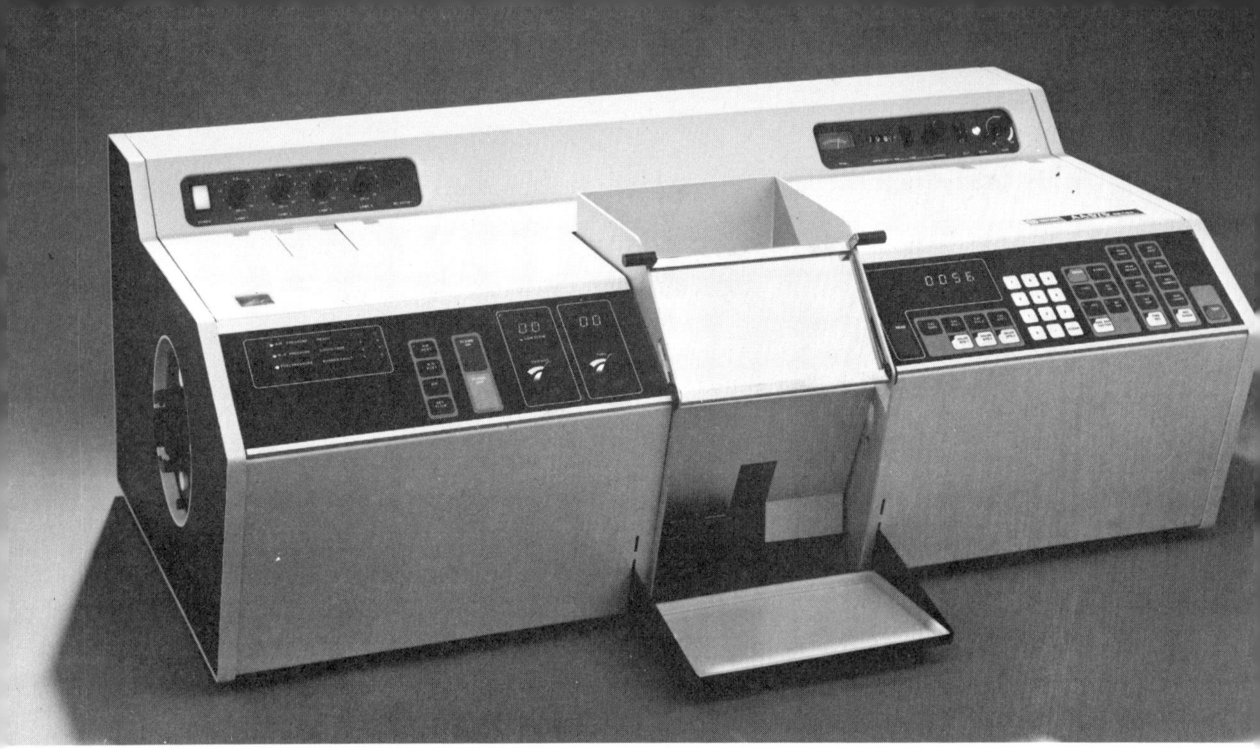

A recent atomic absorption spectrophotometer, with a micro-processor that does nearly all the thinking for the operator.

the light that had passed through the sample and thus showed how much had been absorbed.

Walsh and his colleagues have constantly improved the atomic absorption spectrophotometer since then and it has become accepted around the world, with more than 30 000 units in use in hospitals, factories and laboratories. They are used to measure traces of metallic elements in a whole range of substances, such as soil, blood, urine, plants, minerals, wine and engine oil.

One special improvement was the invention of the carbon rod atomizer, which allows solid samples to be analysed direct, thus saving the time needed to dissolve them.

The wavelengths in use in atomic absorption spectroscopy are generally in the so-called ultra-violet region of the light spectrum, invisible to the human eye.

Atomic absorption spectroscopy is now used in almost every field of chemical investigation from mining to medicine.

One Sydney boy owes his life to the technique. A few years ago he lay seriously ill in a Sydney hospital. He had been badly burnt while playing with a tin of petrol. For weeks doctors fought to save his life, but his condition remained serious. He began to have violent convulsions for no apparent reason. Death seemed near.

Then doctors at the Children's Medical Research Foundation began a series of tests by atomic absorption spectroscopy. They showed that the boy had suffered a severe loss of magnesium in his blood as a result of his burns. Doctors replaced his lost magnesium, the convulsions ceased and eventually the boy recovered.

Measurement of magnesium in blood samples had rarely been attempted because it was so difficult. But with atomic absorption spectroscopy this measurement can be made quickly, simply and accurately.

In pollution control work, the atomic absorption spectrophotometer is used to detect and monitor pollutants in blood and body tissues, and in products such as rice, flour and seafood. It can be used to analyse stream water and trace pollutants to their source. In Canada, it played a significant role in the detection of harmful lead intake by children living near a lead smelter. In such investigations as this, minute blood samples can be analysed rapidly and accurately.

38

Frank Cotton's ergometers helped him pick champions

WITH THE HELP of his ergometers, Sydney University Physiology Professor Frank Cotton could pick potential athletic champions even though they had never shown any great form or even interest in the sport concerned.

Peter Evatt, for instance, had never rowed in his life until one day in 1950 when he met Cotton at a Health Week exhibition in Sydney Town Hall. Evatt tried his strength against the ergometers and Cotton declared he had the makings of a champion rower. The tall, lean 28-year-old lawyer took up the sport, and within two years he was New South Wales and Australian sculling champion.

Frank Cotton was born in Sydney in 1890. The son of a drover who became the State member for Newtown, he went to Sydney High School and the University of Sydney, where he graduated in science in 1912.

At the university Cotton began in 1909 to work out his theories of how scientific principles could be used to make sporting champions. He applied them to his own swimming and set university records for every distance from 220 yards to one mile.

It was not until 1946, by which time he was a professor, that Cotton devised and built various machines he called ergometers. They were nothing more than stationary bicycles and rowing boats with adjustable tensions for users to match themselves against. Cotton used the ergometer findings to decide what training individual athletes needed.

He soon had some impressive results to show. Future Olympic swimmer Denise Spencer used a programme Cotton drew up for her and won the 1948 Australian 440 yards title in 4 mins 31 sec, which was 16 seconds better than she had ever swum before. Edwin Carr won the 1949 Australian 440 yards swimming championship under Cotton's tutelage. Olympic gold medallist John Hendricks consistently followed Cotton's advice.

In 1949 Cotton tested 300 undergraduates at Sydney University on his ergometers and picked the eight-oar rowing team that won the New South Wales title for Sydney University for the first time in 40 years.

On another occasion he picked four rowers for an attempt on the New South

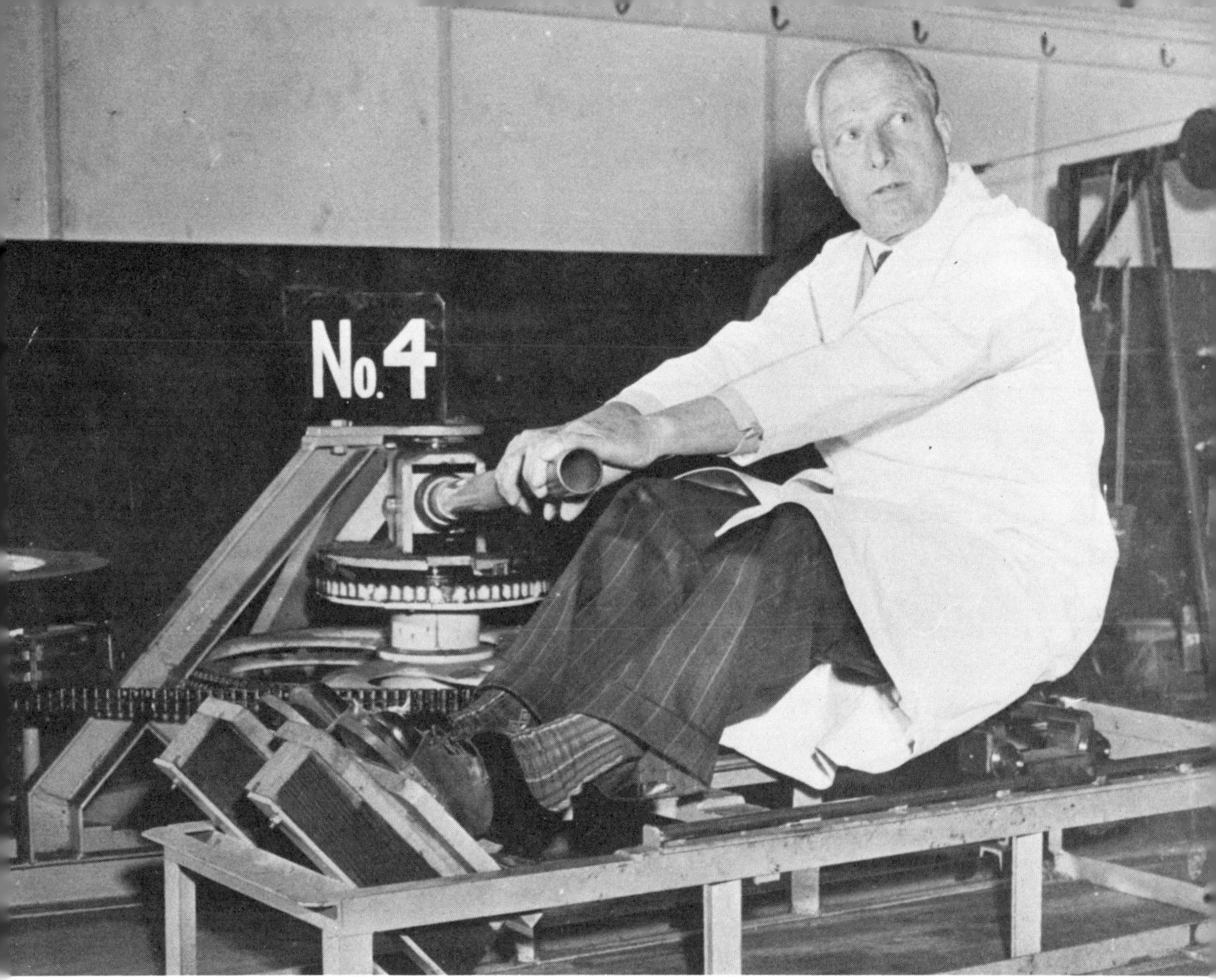

Professor Frank Cotton with one of his ergometers, a rowing machine.

Wales championship. Two of them had never been out in a boat before, but the ergometer was right about their latent ability, for after only five races Cotton's crew won the State title by five lengths.

When Cotton retired from his University duties in 1955, he seemed set to play a big role in preparing Australia's team for the 1956 Olympic Games in Melbourne, but he died only a few months later. Still, many Australian athletes who took part in those games owed their success to Frank Cotton's methods and his ergometers.

39

Vicrail's bogie exchange ended waggon swapping

AUSTRALIA'S RAILWAY gauge problems all started off like an Irish joke, but they certainly weren't funny.

Australia's first railway engineer, one Shields, was Irish, so he decided to use the Irish gauge of 5 ft 3 in for the Sydney to Parramatta railway, instead of the English Standard gauge of 4 ft 8½ in. To avoid any future break of gauge, railway builders in Victoria and South Australia agreed to go along with New South Wales and use the 5 ft 3 in gauge.

But then the Sydney Railway Company reduced Shields' pay, and he resigned. The new engineer, a Scot, decided to build the Sydney to Parramatta line to the 4 ft 8½ in gauge after all. Victoria had already ordered its locomotives and rolling stock, so declined to change its proposed gauge. South Australia decided to keep the same gauge as Victoria.

When Queensland built its first railway, it also had an Irish engineer, Abram Fitzgibbon. He persuaded the Queensland Parliament to build its railways to still another gauge, 3 ft 6 in.

So Australia built its railways to three different gauges, and several places became notable as break of gauge towns. Albury, Wallangarra, Port Augusta, Kalgoorlie, Port Pirie and Peterborough developed substantial local industries just transferring freight from one system to the other.

The worst of the problems disappeared in the 1960s with the building of standard gauge lines from Wodonga to Melbourne, Broken Hill to Port Pirie, and Kalgoorlie to Fremantle. Freight could move from Brisbane to Fremantle in the same waggon. But most of Victoria and South Australia were still on 5 ft 3 in tracks. This problem was substantially overcome by bogie exchange, pioneered in Australia by Victorian Railways.

Compared with the old method of moving goods from one waggon to another, bogie exchange is quick and simple. Standard gauge rails, specially machined to fit, are set between the broad gauge track.

When a waggon comes in for bogie exchange, four 25-tonne electric jacks lift it off its arrival bogies. These are wheeled out and others of different gauge

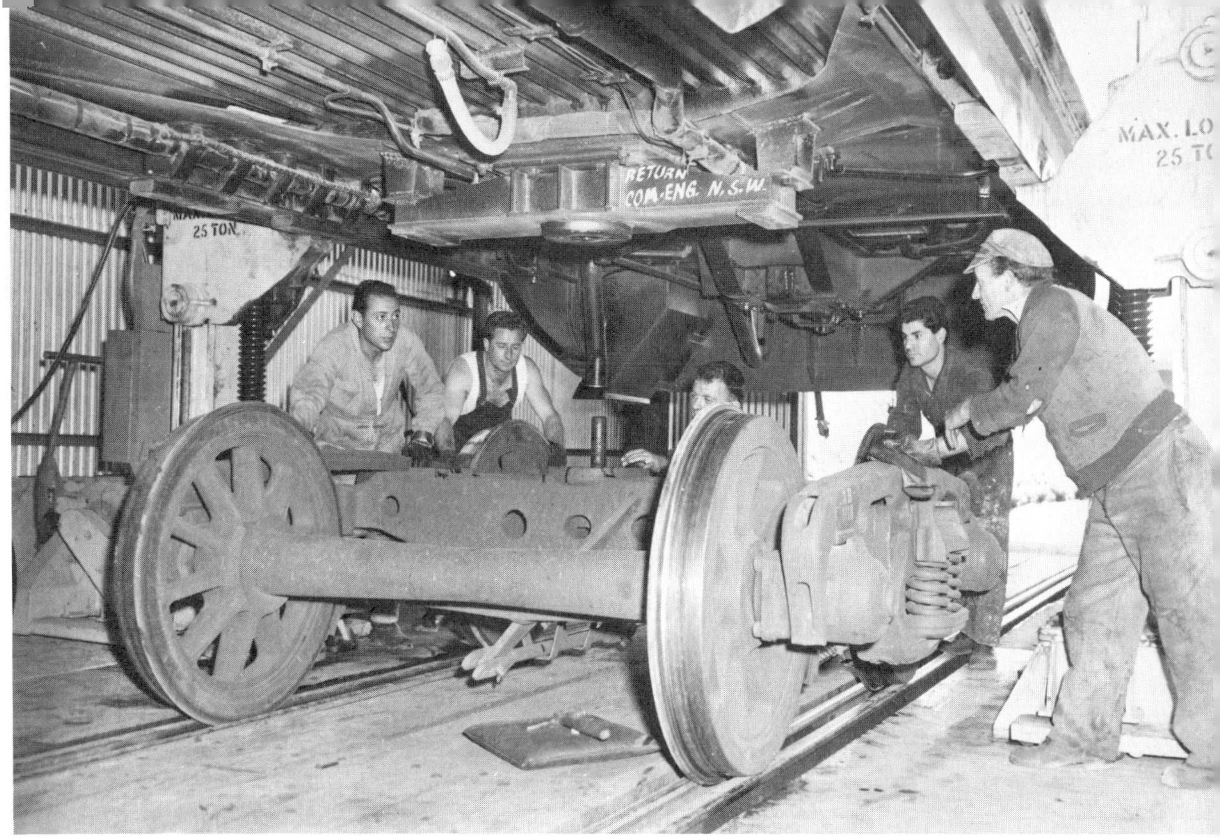

Four jacks lift loaded waggons off their bogies, which are railed away and replaced with bogies of a different gauge.

wheeled under. Then the jacks lower the waggon on to the new bogies, and after final adjustments to such things as brake gear, it is ready to depart on the other gauge. This operation takes about one-twelfth of the time transhipment would need.

Victoria's main bogie exchange centre at South Dynon, covering almost a hectare, was opened in 1963 soon after completion of the standard gauge line to Melbourne. It is at the point where broad gauge lines from the whole of Victoria and from South Australia meet the standard gauge line to all New South Wales and Brisbane. It can handle 80 waggons per eight-hour shift, allowing up to 4 000 tonnes of freight to move undisturbed over both gauges.

Vicrail opened a smaller bogie exchange centre at Wodonga in 1964. This is close to the New South Wales border and caters for freight traffic in north-eastern Victoria.

South Australian Railways in 1965 set up a bogie exchange centre at Port Pirie, where the broad gauge met the standard gauge of the Commonwealth Railways. In 1970 they opened a further bogie exchange at Peterborough, for vehicles handling freight between Sydney and Adelaide.

40

David Henshaw gives an old yarn a new twist

WINNER OF THE 1970 Prince Philip Prize for industrial design, the Australian invented Repco self-twist spinner was the biggest breakthrough in yarn spinning technology since Hargreaves invented the Spinning Jenny in 1764.

David Henshaw conceived the idea of a self-twist spinning machine in 1960 at the CSIRO Division of Textile Industry at Geelong. The new machine spins wool 12 times faster than it had ever been spun before, takes only a fifth of the space, is quieter and more pleasant to work, and uses less power.

Until Henshaw re-thought the whole process, yarn had been made for 4 000 years by continually twisting a strand of fibres in the same direction. This was done by spinning the bobbin on which it was wound.

Henshaw decided to replace this kind of spinning with a new method. He and his colleagues ran two strands of wool through two rollers, in much the same way as clothes go through a washing machine wringer. The rollers rotate but also move rapidly from side to side. This has the effect of twisting the wool

CSIRO physicist David Henshaw produced the biggest advance in textile technology since Hargreaves invented the spinning jenny in 1764.

first one way, then the other. The two twisted strands coming out of the rollers are then brought together and allowed to untwist around each other to form a two-ply yarn.

Once he had validated the new principle, David Henshaw worked with his colleagues Gordon Walls and Basil Ellis to develop a machine that would do it on a commercial scale. They arranged with Repco Limited to build several machines to CSIRO designs. Repco staff helped develop the machine even further, and Repco was licensed to make the new spinning machine commercially.

The Prince Philip Prize for industrial design won by the Repco spinner is organized by the Industrial Design Council of Australia, and aims to recognize products likely to make substantial economic contributions to Australia's advance.

Commenting on the Repco spinner, the judges said: 'The spinning machine is as revolutionary in design as is the process that it employs.

Repco self-twist spinners have production rates 15 times that of ordinary spinners.

'The designers have produced a compact piece of equipment of first-class finish, elegant and aesthetic in appearance, incorporating engineering principles not previously applied in the textile machinery field.

'Particularly notable is the mounting of the self-twist rollers on an air-cushion which allows almost frictionless motion both of rotation and reciprocation.

'Equally ingenious is the actuating mechanism which produces both these motions through the one linkage. In fact, great ingenuity has been displayed.'

One operator can handle a bank of ten machines, equal to about 1 200 spindles in conventional spinning equipment. The machines occupy about 70 per cent less floor space than conventional machines of the same capacity, and need only about 60 per cent of the horsepower. They need considerably less maintenance than conventional spinning machinery.

Two-ply yarn from a Repco spinner. A little uni-directional twisting overcomes the potential problem of the areas of 'no twist' where the rollers have changed direction.

The two-ply yarns produced by the basic self-twist spinning process were more suitable for weaving than for knitting, which calls for greater strength. CSIRO scientists overcame this by wrapping two continuous synthetic strands around a core of wool.

To produce this thread, a new but complementary spinning machine, again developed by the CSIRO in collaboration with Repco, was unveiled at the International Textile Machine Exhibition in Milan, Italy, in 1975. Called the SELFIL spinner, it produces yarn with all the advantages of wool. It is soft, warm and does not snag. Because of its 20 per cent synthetic content it is stronger and more resistant to abrasion and pilling, has fewer knitting faults, better stability, and is easier to cut and make up.

In 1975 the SELFIL Spinner, like the Repco Spinner, won the Prince Philip Prize for industrial design.

41

Burt Terry's Rowcar gives new mobility to handicapped children

MANY HANDICAPPED CHILDREN can look forward to extra mobility, thanks to Burt Terry's Australian designed and produced Rowcar.

At first glance the Rowcar might look like any ordinary pedal car for children. But a closer look soon shows that it has no pedals. Both power and steering are done entirely by the hands and shoulders. For children who do not have the use of their legs, the Rowcar means a new dimension in mobility.

The story of the Rowcar began in 1971 when Owen Eason spoke to Harry Grover, editor of the *Nunawading Gazette*, a Melbourne suburban newspaper. Did Grover know of anyone who could help build a self-propelled car for children afflicted with spina bifida?

Spina bifida, Eason explained, means 'spine split in two', a condition caused when one or more vertebrae fail to close fully before the child is born. This results in loss of control of the legs, feet, bladder and bowels that ranges from mild weakness to total paralysis. Eason knew, because one of his children was born with it.

Modern surgery and specialized medical teams such as the one at the Royal Children's Hospital in Melbourne save most of the infants who a generation ago would not have survived. But survival brings its own problems, notably the child's lack of mobility.

Spina bifida children who cannot use their legs find conventional pedal-driven tricycles and toy cars quite useless. Their lack of mobility and their inability to play like normal children deepens their sense of isolation, making the problems of adjusting to their condition even more critical.

Owen Eason had come to realize that what these children needed was a vehicle they could both power and steer with their arms and shoulders. But nobody seemed to know how to make one.

Harry Grover took the idea to the Rotary Club of Nunawading. Members were soon deeply involved in the problems of mobility for spina bifida children. Past president Phil Gude's enquiries showed that spastic children as well as polio and road accident victims could also use the kind of vehicle Eason proposed.

Another past president, Burt Terry, an engineer who holds patents on refrigeration and air conditioning components, worked out how Owen Eason's basic idea could be made a mechanical reality.

Propulsion could come from a rowing bar in much the same way as it does to the railway handcar. The power could be transferred to the back wheels by a link to a crank. Steering from the same handles could be by a cable linkage to the front wheels.

Rowcar inventor Burt Terry.

Burt Terry built a model and tested it. When he found the model worked, he went on to a full size chassis. The mechanics of the Rowcar had been invented.

Three Nunawading Rotarians—boat-builder Graham Williams, confectioner John Almond and architect Walter Colman—worked out how to mould a body in fibreglass for the Rowcar. They gave it a dished, self-centring seat that would cater for a wide range of body sizes, and two dished areas in front to support the user's legs and protect them from the moving rowing bar.

Volunteer workers turned out parts for the Rowcar and prototype cars were soon available for testing. The teething problems were sorted out and in mid-1972 Nunawading Rotary presented the first batch of 20 cars to the Spina Bifida Association of Victoria.

George Koudsi in Rowcar at Northcott Handicapped Children's School, Parramatta.

The Rowcar was soon such a huge success that Nunawading Rotary had trouble funding it. So Ken Ince, a tailor, organized club members into cutting, assembling and painting more than 4 000 bird feeders which sold briskly through the mail and through club members' businesses. Money from the bird feeders funded about 300 Rowcars.

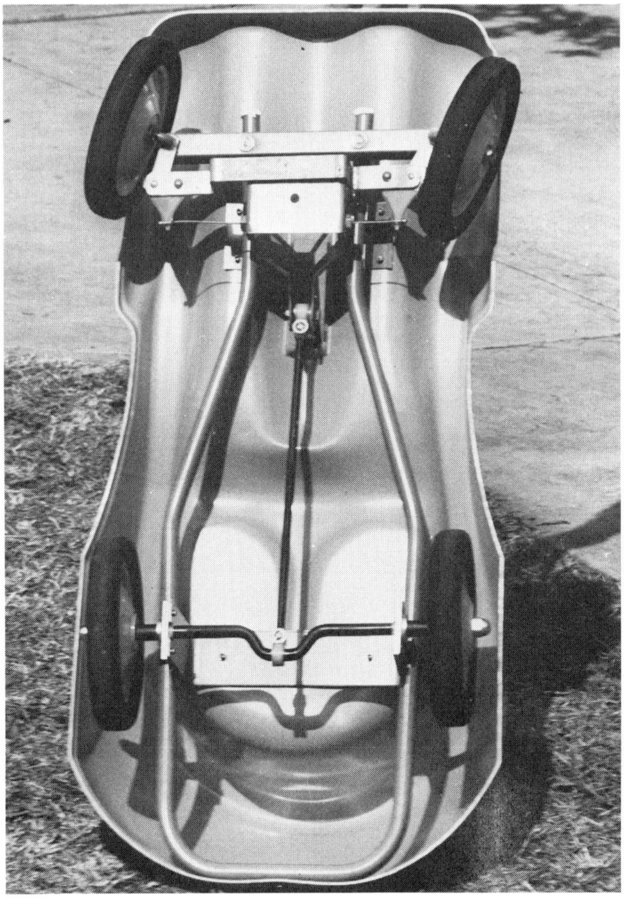

Combined steering and power control of the Rowcar, all worked by hand.

From there the Rowcar went on to win Rotary International District 280's Significant Achievement Award. Demand became so great that the Rotarians could no longer cope with it. The job of mass production was taken over by long-established Australian wheeled toy manufacturers, Cyclops Industries Pty Ltd.

42

Bill Kerruish cuts the work, dirt and danger along with the trees

LIKE MOST FOREST jobs, thinning out pine plantations can be hard, dirty and dangerous work. The Windsor Harvester takes most of the work, dirt and danger out of the job, and cuts costs too.

Bill Kerruish led the research group that since 1970 has developed the Windsor Harvester at the CSIRO Division of Forest Research in Canberra. They worked closely with APM Forests Pty Ltd and Bob Windsor of R. L. Windsor and Son Pty Ltd, who are now making the harvester commercially.

When the Windsor Harvester goes to work, it uses a set of massive hydraulic secateurs mounted on an articulated boom to snip the tree off at the stump. Then its shearing head grips the severed tree while the boom lifts it and topples it backwards, under control, into a bed on top of the harvester.

Bill Kerruish.

The Windsor harvester clamps a pine tree in readiness for shearing it off.

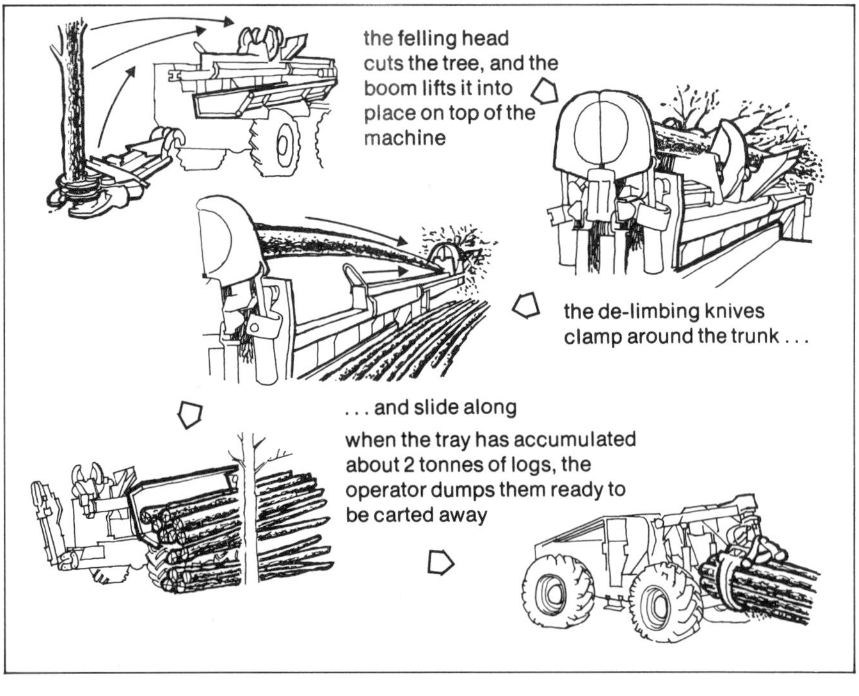

How the Windsor harvester works.

Once the felled tree is in place, two curved blades encircle the trunk and slide along, slicing off tne limbs as they go. Finally, a shear swings down and lops off the top of the tree.

The harvester then rolls the log into a basket on the side of the frame. When the basket contains about twelve tree-length logs, the operator pulls a lever and drops them on the ground ready to be taken to the roadside as a bundle. Through all these operations the driver never leaves his protected cabin.

43

Ralph Sarich puts the internal combustion engine into orbit

LIKE MOST INVENTORS, Ralph Sarich almost lives with his invention, a new concept in internal combustion engines that could transform the industry. But at least Sarich is spared the financial stress that had plagued many inventors. BHP has undertaken to fund his development work.

As a boy, Sarich never took much interest in his Yugoslav parents' orchard at Herne Hill, 30 km from Perth.

'I suppose I was nine or ten when I really started on engines,' he says. 'I used to repair the farm equipment or just poke around. I think I learned more from our tractor than anything. Once I pulled down the hydraulic system to see how it worked, and I remember my father saying that he would need to have the tractor ready next day. I sweated all night over putting it back together.'

After he left school Ralph Sarich qualified as a fitter and turner and studied mathematics, physics, chemistry and engineering subjects at night school. Without bothering about other formal qualifications he soon formed his own engineering company. As well as run-of-the-mill engineering jobs the company developed new agricultural equipment: an earthmoving scoop and an orchard spray.

But Sarich's big fascination was engines. 'They are like a heart' he says. 'They fascinate everyone, even women.' He was surprised to hear an automotive company representative say there was little future for the rotary engine. He felt sure the technical difficulties the car companies saw in existing rotary engines could be overcome.

Sarich started by studying all the registered designs of rotary engines right back through the nineteenth century. Most of these had similar inherent problems which made them impractical. But from this research Sarich soon drew up a list of qualities an ideal rotary engine should have.

How did the Sarich orbital principle emerge? 'The idea just came to me,' Sarich says. 'Then we wanted to try it out as soon as we could.

'We spent three months on the drawings before we built our first prototype, a two-stroke. We finished it in mid-1971. It fired but would not run on its own power, so we started again.

Ralph Sarich (right) with Robert G. Ward, General Manager of BHP's Research and New Technology

'This time we built a second two-stroke. It worked but we were not happy with it.

'It was another year before the third model was ready. This was a four-stroke with six chambers. One Sunday morning in June 1972 it fired for the first time and turned over at around 1 500 rpm for about 20 or 30 seconds.'

There was great relief and excitement in the tiny Sarich workshop that day. Sarich and his assistant, Bruce Fairclough, had spent three months adjusting the timing of the engine's firing. If they had struck more problems at this stage who knows where they would have gone from there?

Ian Miller of the Royal Automobile Club of Western Australia dropped in that day. He had taken an interest in the project from its early stages. Sarich started up the engine again so Miller could see it working. He seemed happier than the whole research team.

They started it a third time just to prove it would start on demand. The flywheel dowel pins sheared with the acceleration but nobody was too worried. This engine still works and has never failed to run.

Work now began on a fourth prototype. It had seven combustion chambers instead of six, so it could have an even firing sequence. This new engine fired on 31 March 1973 with highly satisfactory results. The engine had excellent compression, which showed that its sealing system was working. It ran at speeds of up to 3 800 rpm, at which its operation was quite smooth.

Even before the successful firing of the third prototype, Ralph Sarich and his orbital design had aroused widespread interest in the motor industry. Experts saw that it had great potential.

The engine's first advantage is that it is small, no bigger in fact than the average car's spare tyre. It should be easy to make it on a large scale. It should have a substantially higher power to weight ratio than the normal reciprocating engine. It has fewer moving parts than conventional engines, and because of its simple design wear should be less. Subsequent tests have shown that it runs smoothly and quietly.

The engine's rubbing speeds between moving parts are lower than almost any conventional engine of comparable power, so it should last longer.

Front of a six-chambered four-stroke Sarich orbital engine.

Seven-chambered Sarich orbital engine, the fourth prototype designed and developed by Ralph Sarich.

Sarich thinks the power of the engine can be increased by making it bigger, and the engines can be banked up in the same way as the present Wankel units in cars. But there is no real need for this in cars, since it is estimated that the compact seven chamber prototype now on the benches has the power of a conventional V8 reciprocating engine.

Sarich appeared with his engine on *The Inventors* in September 1973 and immediately aroused the interest of two BHP research executives, Dr R. G. Ward, General Manager, Planning and Research, and Dr H. W. Hosking, Executive Officer, Research. Within a few days the two men were in Perth talking to Sarich.

Later in the year, BHP, Sarich and two of his colleagues had set up the Orbital Engine Co. Pty Ltd to develop the engine. BHP and the Sarich group (Sarich, Tony Constantine, Roy Young) would have equal shares in the company, but BHP would provide whatever money was needed to develop and manufacture the engine.

The greatest immediate potential for the orbital engine is for cars and trucks and, because of its many favourable features, it has already attracted the notice of the big three (Ford, Chrysler and General Motors) and others including Nissan and Renault. Sarich has had several offers of vehicles to test the engines. Other important uses could be in lawnmowers, small boats and light aeroplanes.

The size of the engine will allow car designers greater flexibility in body design and make it easier for them to incorporate such safety features as collapsible front ends, and such luxury items as power steering or air conditioning.

44

Leo Port takes the worry out of catching a lift

Leo port's Port-El Lift Control System takes the work and worry out of catching lifts in modern multi-storey buildings.

In most lift systems you press a button to show whether you want to travel up or down, then wait anxiously to find out which of several lifts will offer to take you, rush to it before the door shuts in your face, then press another button to tell the lift where you want to go.

All of this is just as bothersome to the lifts, Leo Port says. Since no lift knows until you get in just where you want to go, several lifts may stop at the

With the Port-El system passengers do not press an 'up' or 'down' button, but one for the floor to which they are bound.

A panel over one lift indicates it will stop at the floor requested. Passengers can move to the appropriate lift before it arrives.

same floor. And since a surprising number of people press both the up and down buttons in the mistaken belief that this will get a lift to them more quickly, lifts spend a lot of time running empty to pick up passengers who have already left.

In the Port-El Lift Control System, the floor selector panel is not inside each lift but in the lift lobby on each floor. The intending passenger presses the button for the floor he wants. An indicator panel shows which particular lift will take the passenger to the floor he wants. The passenger can move calmly to that lift and wait for it, instead of playing a cat and mouse game with lights, gongs and lift doors. Inside the lift an indicator panel shows which floor the lift is at.

With the Port-El system elevator lobbies can be considerably longer which gives greater planning freedom and can save floor space. Passengers going to a particular floor are grouped into one lift, which minimizes the number of stops lifts must make. This cuts waiting time. The Port-El system can be worked by one of today's mini-computers at no extra overall cost.

Lifts using the system have been working successfully since 1968 at the 13-storey Sydney University Law School and the 11-storey New South Wales Milk Board. The Port Elevator System was awarded the Prince Philip Prize for Australian Design in 1969.

45

Sirotherm: A 'plum pudding' that makes bad water come good

In australia water is a precious resource. Even when it is available it can be brackish, too salty for many uses, especially in machinery. This is often a problem with underground water in particular. So it was not too surprising that the CSIRO should turn its attention to finding a way to separate water and its salty impurities.

Dr Don Weiss has been working on the task since the mid-1950s. A study of Australia's desalination requirements at the time showed that the major need was for a process capable of large-scale but cheap removal of small amounts of salinity from brackish waters or effluents.

Most established processes removed the larger volume, the water, from the smaller volume, the salt. Distillation operates in this way, and uses large quantities of heat energy. The reverse osmosis process also removes the water from the salt, and uses large quantities of mechanical energy.

Don Weiss and his team felt they should concentrate on finding a way to remove the comparatively low volume of salt from the water. They knew that this could be done by ion exchange, in which some suitable substance attracted the positively charged particles (ions) which comprised the salty impurities. But this method required expensive chemicals to reverse the ionization process and restore the original ionizing medium for further use. What Weiss needed was an ionizing medium that enabled the process to be reversed and the medium restored cheaply, such as by applying heat or electricity.

In their first approach to the problem Weiss and his team used carbon electrodes, but they quickly gave them up in favour of ion-exchange polymer resins. These would remove the ions from cold water but could be re-used after flushing out the trapped impurities with relatively inexpensive warm water.

Years of work went into studying the most suitable resins, then synthesizing and testing them. The CSIRO researchers found resin combinations that showed a large, thermally-dependent ability to absorb salt.

Then they moved on to finding ways to increase the rate of salt absorption of these resins. They concentrated on using micron-sized resins since such finely-

A laboratory testing rig for evaluating Sirotherm applications.

divided resins react very rapidly. They developed the 'plum pudding' concept, in which micron-sized resins (the plums) are mixed within a porous inert matrix (the pudding).

By 1968 Don Weiss and the CSIRO workers had a laboratory column in operation, using 'plum pudding' resins and working at commercial flow rates.

The next step was to advertise for a collaborator to get the new idea into commercial production. ICI Australia Ltd became co-developer of the CSIRO

Sirotherm plant under construction at ICI plant, Osborne, South Australia.

patents. Years of further research followed, with pilot plants eventually proving the system's worth. One plant removed brackishness from bore water in Perth. Others desalted sewage effluents in Melbourne and later in Tokyo.

The name Sirotherm was derived partly from CSIRO, partly from 'therm', for the heat used in the process.

The first Sirotherm plant on a commercial scale began operating early in 1976 at the ICI works at Osborne Park in South Australia. It treats Adelaide tap water to reduce its salinity from 600 mg/l to less than 100 mg/l before it is fed to a conventional demineralization plant. This results in a considerable saving. The treated water is used in a boiler.

Planned Sirotherm installations include one in Bangkok for demineralizing water for boilers, one in Japan for re-use of process water in a chemical plant, and one in Okinawa for the sugar industry.

46

Interscan brings the world's aircraft down safely

AIRPORTS ALL OVER the world could soon be switching to Interscan, an Australian designed system that enables aircraft to approach at a steeper angle and on a curved path, giving faster landings and fewer noise problems.

In the previous instrument landing system, planes came down a three degree glide path to a touchdown along two radio beams, one giving elevation, the other giving azimuth, or line of approach.

In 1968 the International Civil Aviation Organization began to think about a more advanced landing system for the future. In some areas, notably North America, a new system was already overdue, because of increased traffic density.

The Organization in 1972 specified what would be needed in the new system. It invited member states to submit proposals for systems to meet these needs, on the understanding that their governments would develop their proposals and produce hardware and flight test data for comparative evaluation. Australia entered the competition, along with Britain, the United States, West Germany and France.

The new system would need to allow an aeroplane to fix its position with great precision in three dimensions: within 40 degrees of either side of the centre of the runway in azimuth, up to 20 degrees in elevation, and out to a distance of 30 nautical miles. It would also need to allow the aeroplane to make a curved approach of any specified form within this sector, then touch down in all weathers even in zero visiblity and without the aid of a radio altimeter. It also had to provide continued guidance for any plane that missed its approach and flew on.

A committee in the United States proposed that five transmitters on the ground should be used to provide the landing system, and there was general acceptance of this. One would be the azimuth transmitter to give the aeroplane lateral guidance on either side of the runway. A second would be the elevation transmitter, to cover angles from zero to 15 degrees. A third would cover the back azimuth for missed approaches. A fourth would give very accurate height information in the final 45 metres of descent. The fifth would give accurate distance measurements by radar.

Australia's CSIRO Division of Radiophysics in co-operation with the then Department of Civil Aviation became involved in the project, and soon began to feel it had better ideas than those being proposed by other countries.

CSIRO researchers led by division chief Paul Wild felt the best way to give an aeroplane data on its angular position was a radio beam wide (20 degrees) in elevation, but very narrow (1 degree) in azimuth, and to scan this beam at a set rate from left to right across the 40 degrees on either side of the runway. The receiver in the aeroplane would detect two pulses as the beam swept across it. Time between the pulses gave a direct measure of the angular position.

This time measure was the crux of Interscan, originally the only method to use it. Time was far easier to measure than frequency, upon which the other systems depended. On the ground it called for only one transmitter emitting one frequency.

To work out this concept was substantial progress, but it still had to be put into practice, and in short time too. CSIRO people went on to produce basic designs and experimental equipment. Amalgamated Wireless Australia produced prototypes of the actual equipment that would be used in the field. The Department of Transport provided the financial and contractual administration, the data necessary for operational input, and the flight-test organization.

When Interscan was demonstrated to an International Civil Aviation Organization working group in 1975 at Melbourne airport, the pilot put his aircraft down only 60 centimetres from the centre of the runway.

Experimental model Interscan azimuthal aerial at CSIRO Radiophysics Division at Epping, NSW.

Original Interscan elevation aerial installed at Melbourne Airport (Tullamarine) for flight testing.

In developing Interscan, the CSIRO Division of Radiophysics was able to fall back on its considerable experience of aircraft guidance systems. This had started during World War II, when overseas systems were adapted for Australian use. It led in 1945 to the development by Brian Cooper of distance measuring equipment for civilian aircraft.

In distance-measuring equipment, a transmitter in the aircraft send signals to a receiver on the ground, which responds with a reply pulse. The equipment measures the split section of time between sending the original signal and receiving the reply, uses this to compute the distance, and shows the result on a graduated distance indicator in the aircraft flight deck.

Australia's distance measuring equipment operated from the beginning on the 200-megahertz band, using well-established techniques, which meant it could be developed rapidly. The first trials were conducted in 1945, and by 1947 two commercial aircraft were carrying distance measuring equipment on their regular runs between Sydney and Melbourne. The equipment was so successful that the Department of Civil Aviation set up about 100 unattended, automatic ground beacons to respond to aircraft signals. The system was fully operational by 1953, and within four years all Australian domestic airliners were fitted with distance measuring equipment. In 1956 the job of converting all distance measuring equipment from valves to transistors began.

In contrast to this, the 1000-megahertz distance measuring equipment network in the United States was not operating until 1962.

47

'Pure serendipity' in electrical measurements for Mel Thompson and Doug Lampard

MEL THOMPSON modestly describes as 'pure serendipity' the discovery he shared with Doug Lampard of vastly improved methods of setting electrical standards of measurement, methods which have made Australia the world's leading nation in that field.

'Serendipity' is making happy and unexpected discoveries by accident. Perhaps Thompson is a little too modest. The discoveries may have been happy and unexpected, but they were hardly by accident. The two men had been working with others on the problem of electrical measurements for some time.

Every year, Australian firms make millions of dollars worth of electrical equipment. To ensure that it performs well, they need constant checks and rechecks of the basic units of electrical measurement, the volt and the ohm. The electrical values of such items as capacitors and resistors used in electronic equipment must also be checked and kept accurate.

The long accepted way to set standards of electrical measurements was to make a coil of copper wire of a large number of turns wound on a very precise former, usually of fused silica. By measuring the position and diameter of each turn of wire, experts can calculate the coil's inductance. This is a measure of the magnetic energy generated when a current is passed through the coil, and depends only on its geometry.

Once the inductance of the coil is known, a complex series of electrical measurements can be used to relate the resistance of a resistor to the inductance of the coil. With the absolute value of one resistor established in this way, it becomes the standard and the value of other resistors can be found by comparing them with it.

The bugbear is the measurement of the coil. It is a finicky and frustrating business that can take up to six months.

In the early 1950s the CSIRO's National Standards Laboratory in Sydney worked out how to make exact measurements of small values of electrical capacitance, and to compare capacitance and resistance. By 1955 scientist Keith Clothier had a well-developed design for a variable capacitor of two metal-coated flat glass plates in a vacuum. But even this required measurement of the plates and their spacing to get an accurate value.

At this stage Mel Thompson decided that the measuring job would be easier if the capacitor was a simple cylinder divided into two pairs of electrodes. Any imperfections in the symmetry would be self-cancelling. Doug Lampard calculated the capacitance of several different cross-sections and proved that it depended only on length and not on size or shape of cross-section. This was entirely unexpected, but tremendously useful, the 'pure serendipity' that Thompson spoke about.

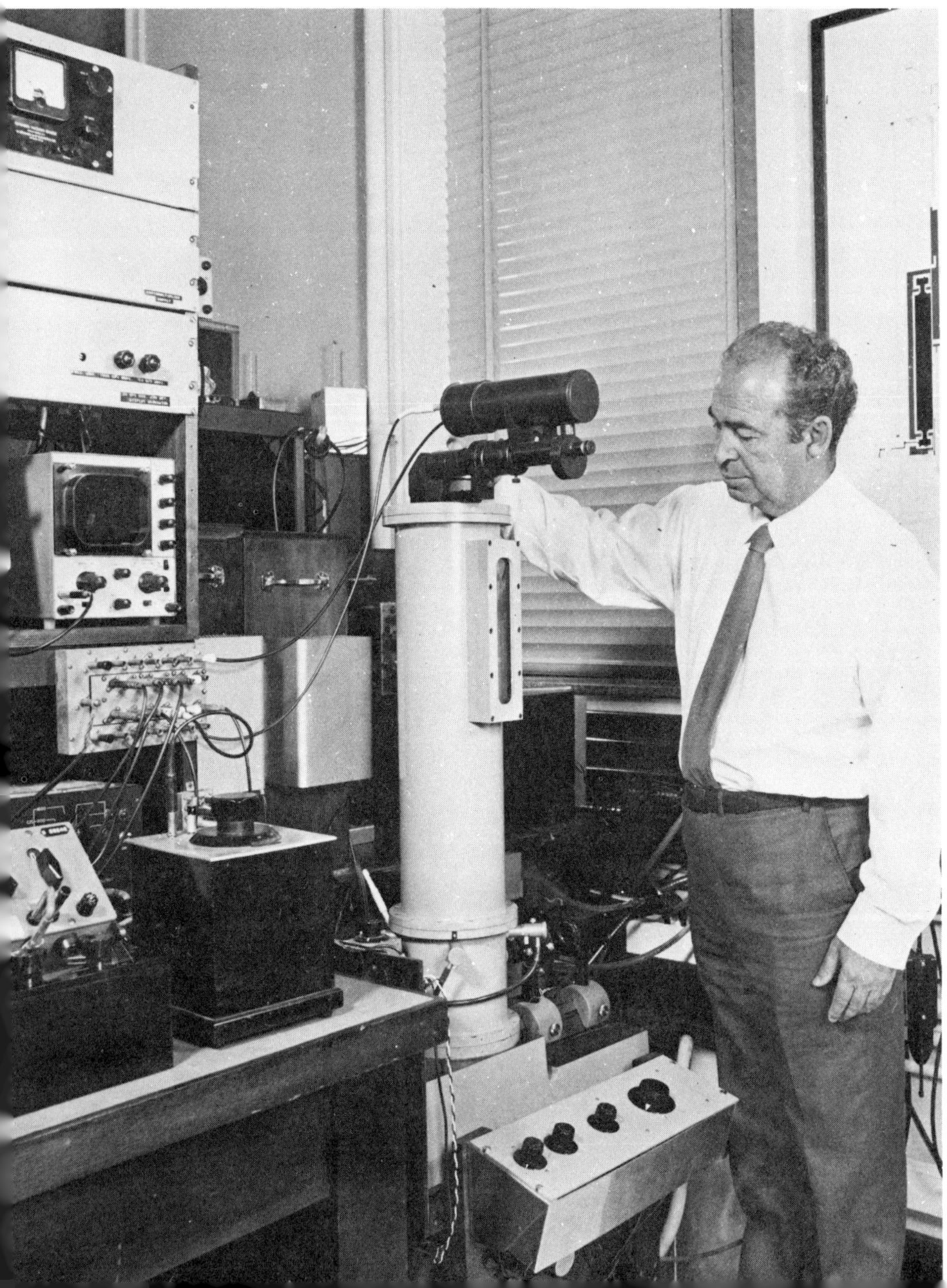

Mel Thompson with the 'Calculable capacitor'.

The 'Calculable capacitor'.

In co-operation with Clothier, Thompson and Lampard worked towards a capacitor of four parallel brass rods whose capacitance could be calculated from one length measurement. This had to be very exact, and was done with light waves. The scientists expected an accuracy of one part per million, but to their amazement achieved an accuracy ten times that. Later Thompson said they could get an accuracy ten times better still.

From 1964 to 1974, when the Americans confirmed the value with similar equipment based on Thompson's idea, the National Measurement Laboratory gave the world its best value of the absolute ohm, the measure of electrical resistance, and readings that took not six months but three hours.

During this project Thompson found other important applications for the accurate measurement of small capacitances. These included greatly increased accuracy in the measurement of magnetostriction, thermal expansion, and absolute pressure.

48

Leo Port's Geometrik clubs make life easier for golfers

Leo port admits that perhaps life wasn't meant to be easy, but he denies that golf was meant to make it harder. 'Golf should take our minds away from life's problems,' Leo says, 'not add to them.'

So as an incentive to golfers Leo invented the Geometrik golf clubs. The geometry of the club has been modified so hands point at the 'sweet spot' on the club face. This gives Geometrik clubs an inbuilt guidance system which capitalizes on instinctive human abilities of sight and muscular co-ordination.

The new club first appeared on *The Inventors* on television in 1975, then underwent two years of refinement by Gold Medal Golf Pty Limited, who had it in production by late 1977. Managing director of Gold Medal, Gene Hoffman, says that unlike many other computer-age clubs, the Geometrik meets all the requirements of the rules of golf.

'The Geometrik looks like a conventional club but the similarity ends there,' he says. 'After only a couple of holes you start hitting the ball with more authority. Everybody who has tried them remarks on the improved feel right through the swing.'

Leo Port with one of his Geometrik golf clubs.

49

BHP solves the ring-pull problem with the push button can

THE RING-PULL soft drink and beer can was a great idea, but people aren't always as careful as they might be about what they do with the ring-pull that made it all possible. Discarded ring-pulls have become one more litter problem all over the world and have killed cattle and wild deer.

Australia's steel giant, BHP, has solved this problem by giving the world the Presto can. Instead of adding to the litter problem, its opening mechanism, a simple press button, stays attached to the can. It is easier to produce, and it uses less material.

The story of the Presto can really began in the early 1960s. Aluminium marketers stole a march on the steel industry when they invented the ring-pull can. Until then, steel had been used almost exclusively in cans. Now one end of almost every can was aluminium.

Still the news was not all bad for the steel makers. Cans with ring-pull tops were so popular that more and more of them were sold, and demand for steel for soft drink and beer cans shot up.

Even so, steel researchers all over the world busied themselves in trying to find a way to beat or at least duplicate the ring-pull principle, using steel instead of aluminium. Researchers at BHP were among them.

Since the world uses 50 000 000 000 beer and soft drink cans a year, the market was worth pursuing. Australia's share of this is about 1 500 000 000.

Although BHP experts had been looking at the idea of a new all-steel easy opening can end for some time, work on the project was spurred by BHP chairman, Sir Ian McLennan, in 1968. He roughed out an early design for a press button end.

Early tests on an all-steel push-button top at BHP's Port Kembla Research Laboratories were promising. So too were tests on other types of easy opening end, including an all steel ring-pull that remained attached to the can. Yet although various methods seemed to work well enough technically, there were some doubts about their ability to win wide consumer acceptance.

In a major rethink of the project in 1970, BHP called in Mike Debenham, of IDR Pty Ltd, a firm of design consultants.

To be successful, it was clear that a new type of can-end would need to satisfy several needs. It should be easy and economical to handle on existing high-speed canning machinery and equally easy and economical to make from existing grades of tinplate and aluminium. Since consumer acceptance was vital, it also should be easy and pleasant to use. And on environmental grounds, it should have no detachable part.

The research group decided to persist with the press-button approach. Three more years of research and testing led to the Presto can-end. One major breakthrough was the development of a metal die that forms the hinge and makes the press-button smaller than the pouring aperture.

The Presto system uses two press-buttons. The pressure inside a can of beer or soft drink can be quite high, and thumb pressure has to be concentrated on a

Pressing the small button releases the gas pressure. Pressing the large button releases the beverage.

"HARRY'S DOING THE LAWN WITH HIS USUAL PUSH-BUTTON PRECISION..."

The Melbourne Herald, Nov. 17th, 1973.

very small area to overcome it. Once the first tiny button has been pushed in and the pressure released, the job of pushing in the pouring hole button is easy. Both buttons are hinged on the can and therefore cannot fall into the drink or off the can to create an additional litter problem.

Once BHP and IDR thought they had the problem solved they started a long process of testing. Cans with Presto ends were filled and stored under various temperatures and conditions for six months and more. They were tested at the same time for performance and durability under all consumer handling conditions, and to see that they conformed with health department requirements.

During 1972 BHP's Melbourne Research Laboratories produced can ends and sent them for trials in major overseas countries including the United Kingdom and United States. In three main studies, consumers showed that they preferred the two-button Presto end over the ring-pull and single aperture press button can-end. They favoured Presto for its ease of opening, safety, ease of drinking from the can, reduced pollution potential, appearance, and ease of pouring.

The Presto was a big breakthrough in can-ends for both steel and aluminium. BHP has licensed several manufacturers to use it, including Toyo Seikan, Japan's biggest can-maker, which makes the Presto can-end in steel.

50

Jack Grant's aircraft slide inflates and floats away

WHEN A JUMBO JET makes an emergency landing, a lot of people want to get out in a hurry. That's why jumbo jets have escape slides that inflate on demand to provide a safe, soft and speedy ride to the nearest piece of ground.

If the emergency landing is in the sea, there's the further problem of staying afloat. The traditional solution to this was inflatable rafts stored in the aircraft ceiling. Getting these out of the aircraft was just one more problem that people could do without in an emergency.

So Jack Grant, Qantas Operations Safety Superintendent, decided to think of a better way. In 1965 he had the idea of combining the escape slide and the life-raft into one inflatable item that got itself out of the aircraft. So nowadays jumbo jet escape slides are in the shape of a raft which can be detached from the aircraft and allowed to float away.

Jack Grant on his combination escape slide and raft in the Qantas crew training centre, Sydney.

Grant put the proposal before aircraft manufacturers in the United States and Britain. The slide-raft's effectiveness had to be demonstrated to airline operators, government agencies, aircraft manufacturers and survival equipment experts of many countries. Sea trials were held in Barbados, Honolulu and Sydney.

Working with the International Air Transport Association, Grant formulated and refined the specification of the slide-raft until it was accepted by the United States Federal Aviation Administration and the British Air Registration Board. The slide-raft is now fitted to Boeing 747, Douglas DC-10, Lockheed 1011 and Concorde aircraft.

Jack Grant won the Cumberland Trophy, one of aviation's premier safety awards, for the idea of the slide-raft. The trophy is awarded by the Guild of Air Pilots and Air Navigators for an outstanding contribution to air safety.

51

Edmund Potter shows how TEA can cure smoking

USE TEA TO cure smoking? The idea seems preposterous. But this is no ordinary TEA, no ordinary smoking. In this case, TEA is triethylamine, a common industrial chemical, and the smoking is not the sort done with cigarettes, but the sort done by industrial chimneys.

Australia's need for TEA goes hand in hand with our wealth in coal which, although an excellent fuel, has one major problem in the residue of fly-ash that remains after it is burnt.

If this is allowed to pass out the chimney it becomes a serious source of air pollution, and later settles on the surrounding area. In most places clean air laws require that most of it be removed from the chimney exhaust.

The usual way to remove fly-ash is to pass the chimney gases through electrostatic precipitators. These give an electrical charge to the small particles as they rush through the equipment. They are then attracted to plates which have been given the opposite charge. From these plates it's a simple matter to deposit the collected particles into hoppers and take them to a suitable disposal area.

Some coals, however, are more troublesome than others. They give a particularly fine fly-ash, which does not fully respond to the treatment it gets in the electrostatic precipitators.

Because of this problem the CSIRO in 1965 stepped up its research into the subject. It designed and constructed the only laboratory-scale combustion and precipitator rig in the world capable of simulating the production and collection of fly-ash as it occurs in power stations. Using this rig, CSIRO scientists gradually gained a much greater understanding of the processes at work in the formation, transport and capture of the particles. They established that only three factors play a major part: gas velocity, electrical resistivity and the size of the finest particles.

The rig showed that the very fine particles are much harder to trap than the larger ones. Particles are normally measured in microns, one micron being a millionth part of a metre. About 99 per cent of particles 10 microns in size can be caught in electrostatic precipitators, but only about 66 per cent of particles five microns in size. Particles only one micron or two in size have a very good chance of slipping through.

Some coals produce fly-ash particles most of which are five microns or less, and these used to cause considerable pollution problems.

A research team headed by Dr Edmund Potter of the CSIRO Division of Mineral Chemistry took on the task of finding ways to get the smallest particles

Edmund Potter.

to join together into bigger particles, so that the electrostatic precipitators could deal with them.

An ammonia additive works in some cases, but not well enough to meet clean air act requirements without building new, larger and very expensive precipitators. So the research team tried many other chemicals, particularly organic compounds containing nitrogen, which seemed to offer the best prospects, because of their molecular structure.

They became particularly interested in triethylamine, or TEA. Added to flue gases at a concentration of only 10 parts per million, it increased the efficiency

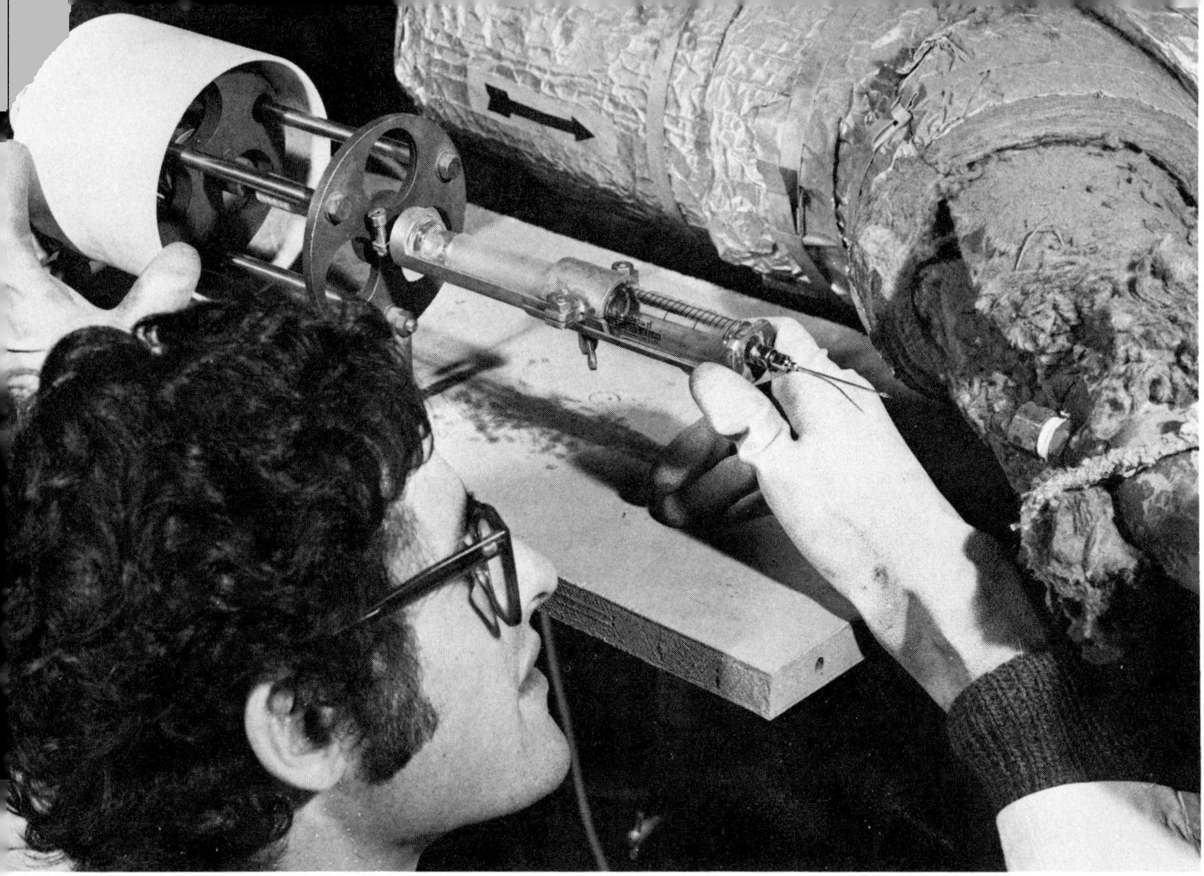

Injecting additive into gas flow.

of precipitation enormously. As well as causing the elusive smaller particles to clump together, it also enabled a stronger electrical field to be maintained in the precipitator without short circuiting. It was 25 times as effective as ammonia.

Proof of TEA's efficacy came when Southern Portland Cement asked the CSIRO for help in soving its smoking problems at its Berrima cement works powerhouse. Coal being used was burning to a very fine fly-ash, only 80 per cent of which was being trapped in the company's precipitators, compared with the legal requirement of 98·5 per cent.

Tests at Berrima showed that 99·7 per cent of the fly-ash could be eliminated from the chimney emissions by adding 20 parts per million of TEA, and that the fly-ash content could be brought within the legal requirements by adding only eight parts per million of TEA.

The company was able to comply with the law without having to build the extra $300 000 worth of precipitators that would have been needed without TEA.

52

Ted Pritchard's steam car can run on alcohol and never pollutes

---

TED PRITCHARD has had his ups and downs since he first decided that steam cars had a future. That was back in the early 1960s, when nobody had heard of the energy crisis or the perils of pollution.

Now, there's great interest in his work. At its present stage of development, the Pritchard steam power unit is far ahead of most internal combustion engines in its fuel economy and its pollution-free emission. So far, it's the only engine in the world capable of driving a motor car and meeting the standards that will apply in the United States in 1985, when a fuel economy of 32 miles per gallon and specified low emissions will be required.

Over the years Pritchard has had a steady stream of people who backed him by taking shares in Pritchard Steam Power Pty Ltd, though at times the money side of the operation has been touch and go. But, thanks mainly to the involvement of the Australian and Victorian Governments, which have put $150 000 into the project, Ted Pritchard is now able to get on with his work without having to worry too much about finance.

In the 1950s, when Pritchard was in his early twenties, he bought his first steam car. It was a Stanley Steamer, probably the best known of the many steam cars that were once on the market. He drove it for years and studied it carefully, learning much about the vagaries of steam cars. Not the least of these was that it took twenty minutes to light the burners and get up steam!

He went over the whole system of steam cars, identifying their weaknesses, and gradually re-designing individual parts to eliminate problems. He and his father designed, built and installed a steam power unit in their three-tonne truck.

Then they built another and installed it in a Ford Falcon. This attracted such attention that some Americans who were backing the venture arranged in 1972 for the car to be flown to the United States.

In a series of demonstrations near Los Angeles, Ted Pritchard showed off the steam powered Falcon to news media, representatives of the state and federal bodies interested in exhaust emissions, to motor car manufacturers including Ford, General Motors, American Motors, Leyland, Volkswagen, Mercedes-Benz,

Ted Pritchard (left) with one of his assistants, Mike Edwards, the Pritchard steam engine, and a prototype of the Pritchard car.

Nissan, Toyota and Alfa Romeo, to John Deere, Aerojet Liquid Rocket, Sierra Club and the United Auto Workers Union.

Some of the comments from the technical people were: 'Boiler control is good'; 'Impressed with driveability'; 'You have a good *little* car there'; 'Impressed with the power'.

Ted Pritchard has put much of his development work into the control of the steam generator, an aspect on which many earlier steam car projects have come to grief. He has Australian and overseas patents on his arrangements for quick response of the thermostat which controls the boiler feedpump and injector.

Ted Pritchard's steam car is not at all choosy about the fuel it runs on. It will accept petrol, but it doesn't insist on it. Kerosene will do. So will diesel. Or, what is more interesting, it will run quite happily on vegetable fuels such as alcohol, which can be made easily and in large quantities from sugar cane, or on vegetable oils. Any fuel derived from crops is, of course, infinitely renewable.

Another possible fuel is low-grade petrol of the kind that can be produced from coal.

The Pritchard engine has low pollution emissions for two reasons. One is that it does not suffer from the knocking problems of the internal combustion engine, so it does not need lead in its fuel. The other is that it gives better combustion. In an internal combustion engine there is a wide range of temperatures inside the combustion chamber. Near the walls it is comparatively cool, so combustion is incomplete, and carbon monoxide forms. Closer to the centre of the combustion chamber it is relatively hot, and this results in the formation of various oxides of nitrogen, particularly as the combustion gases have already been heated by compression. Oxides of nitrogen in car emissions are the major cause of photo-chemical smog. None of these conditions occur in the Pritchard engine.

The two litres of water used to generate steam in the Pritchard power unit are condensed in a radiator and used again. A 16 litre reserve tank is refilled only occasionally. Compare this with the 90 litres water tank of the Stanley Steamer, which had to be filled about every 300 kilometres.

The steam generator supplies steam to drive the engine proper, a two-cylinder 90° vee. Described as a one-stroke, it has steam inlets at both the top and bottom of the cylinders, so that the piston is driven on both the upstroke and downstroke. This particular cycle gives the two-cylinder one-stroke engine just as many power strokes as a conventional eight-cylinder four-stroke engine.

A cutaway diagram of the Pritchard steam car.

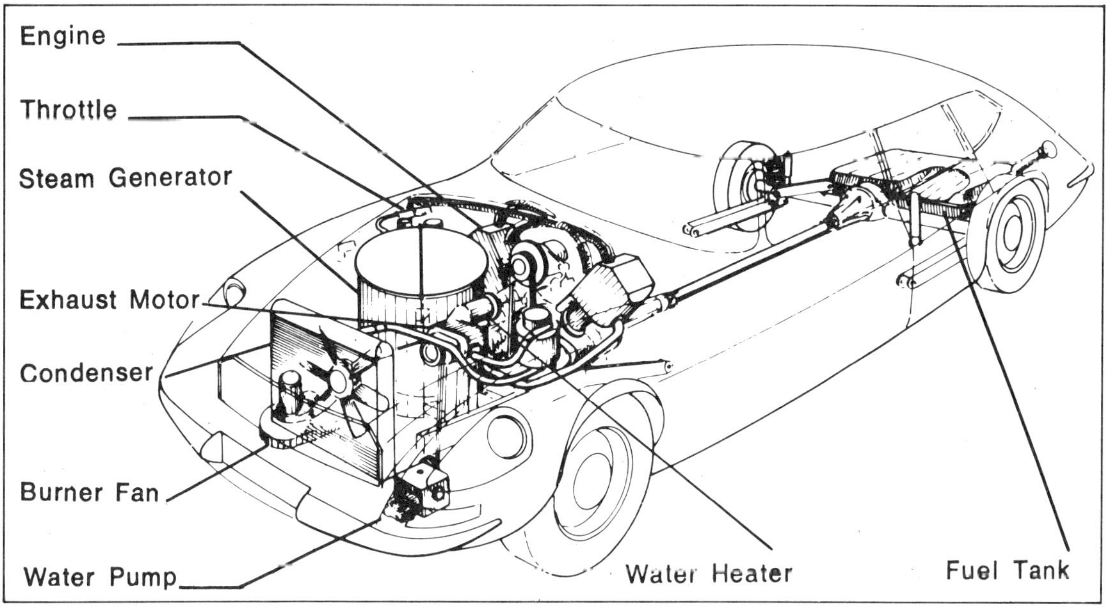

The Pritchard engine has electric ignition for its boiler fuel, which works in much the same way as a domestic oil heater. It takes about 45 seconds to operate from a dead cold start. This certainly is a lot better than the twenty minutes of the Stanley Steamer, but it still seems a long time in comparison with a petrol engine. Until, of course, you remember those times when it took a lot longer than 45 seconds to get started on a cold morning. Once the Pritchard engine has its first warm up for the day, starting is quicker.

Like all steam engines, the Pritchard car engine has high torque at slow speeds, so it does not need a gearbox or automatic transmission. It is easy to drive and manoeuvre.

The Pritchard engine is quiet, which Ted Pritchard says gives it a great advantage for trucks, buses, construction equipment and agricultural machinery. Many plant operators and farmers have their hearing affected by noisy machinery.

The Pritchard steam engine offers safety too. The steam generator consists of coiled tubing which in a heavy front-end collision acts as an impact absorbing crushable structure, unlike the heavy cast-iron block of an internal combustion engine. There are no pressure vessels to cause explosions. If the brakes fail, the driver can simply slip the engine into reverse and bring the car to a halt.

In 1978, Ted Pritchard was working on three prototype engines that were made at the Bendigo Ordnance Factory to test their long term durability. After six months of bench testing, one was to be installed in a specially designed aerodynamic body for twelve months of road testing. A second would then go on the bench for testing, while the third was held as a spare.

Once the engines have been put through their test programmes, Ted Pritchard has plans for a small production run.

53

Storehouse of a thousand inventions

AUSTRALIANS SEEM to have just as much talent as ever for inventing.

Every year, when producer Beverley Gledhill sets out to plan another twenty or so episodes of *The Inventors*, she has a bank of over 1000 inventions to choose from. Of these, she'll need about 80.

Beverley Gledhill has been producer of the top-rating television programme since the Australian Broadcasting Commission first started producing it in Sydney in 1970.

In the show's early days, of course, there was no such bank of inventions to draw on. Beverley Gledhill just had to go out and find them.

'We just went around asking people if they knew any inventors', she says. 'We asked at universities, we asked at technical colleges, and we asked at garages. Many of our best leads came from taxi drivers.

'Once the first programme went on air we had a steady flow of letters from people who had invented something, and since then our major problem has always been one of selection rather than collection.'

Inventions to be used in *The Inventors* are chosen partly on their merit as inventions, but mostly on their likely interest to a mass audience.

'Naturally we look for something with visual appeal, but we aren't hung up on it,' Beverley Gledhill says. 'When we had Ralph Sarich on the programme we found it hard to make his engine look visually interesting, but we felt the invention was of such widespread interest that we could live with that.

'We don't give in to the temptation to use zany inventions just because they are zany. They have to fulfil a genuine need somewhere in the community.

'We don't use very esoteric inventions either, no matter how worthy they are. If somebody invents a special grubscrew that solves some great problem of heavy pressure vessels, we might be prepared to admit it has a great future, but we don't put it on the show.

'We do make something of an exception in the case of inventions related to safety. We happen to know that most people aren't very concerned about safety, not if they have to do something about it or spend money on it, anyway. But in spite of that we do think it's up to us to give some exposure to worthwhile safety inventions.'

The format of *The Inventors* has been polished considerably since its early days, but it still takes the same general form. Compere Geoff Stone introduces each inventor, then shows the invention in action. Panel members Leo Port, Diana Fisher and Vic Nicholson comment on it and award it points so a winner for the night can be chosen.

Leo Port is on the panel in the role of consultant engineer to comment on the working and production practicalities of the invention.

Diana Fisher says her role is 'multi-purpose', that of 'granny, housewife, intelligent university student, inquisitive child, and common sensical, practical down to earth mother'.

Vic Nicholson is a marketing consultant who comments on the likely commercial success of the invention.

Beverley Gledhill says she gets a lot of satisfaction from seeing inventions that first appeared on the show then appear in the shops some time later. She usually loses contact, however, with people who have been on the programme and cannot say whether any of them have ever become wealthy from their invention.

'On-air' personnel of 'The Inventors'. Left to right: Vic Nicholson, marketing consultant; Diana Fisher, 'housewife and mother'; Geoff Stone, compere; Leo Port, engineer.

'If they ask me, I usually advise them not to be greedy. I tell them to go to somebody with the marketing and manufacturing skill, and let them take the risks and supply the resources to get the product on the market.

'We do know that one of the results of the programme has been that inventors now get a much better hearing than they did before it started. Inventors were often

Beverley Gledhill, producer of 'The Inventors'.

treated like cranks and never allowed past somebody's secretary. Now many managing directors' doors are open to any inventor who wants to talk to them.

'We know that many firms nominate people to watch the programme and keep alert for any invention that might have a commercial future in the kind of business they happen to be in.'

All Australians seem to be fairly inventive, but Western Australians and North Queenslanders seem to have more than the national average talent for it, judging by the flow of interesting inventions they send to *The Inventors*.

'Perhaps people in remote areas have more time to think about inventions,' Beverley Gledhill says. 'They don't have to devote so much of their energy to the

Pat Williams, inventor of the split nut which he showed on 'The Inventors' before winning a gold medal at the annual inventions exhibition in Geneva.

problems of big city living. Or perhaps their remoteness means they develop a habit of working out solutions to their own problems.

Two inventions which have appeared on *The Inventors* have gone on to win gold medals at the annual inventions exhibition in Geneva. Jamie Farfor won the first with his ferro-cement building panel, Pat Williams the second for his split nut.

Pat Williams' invention is so simple that almost everybody could wonder why they hadn't thought of it first, but it has very wide potential application.

His split nut is made in two halves, separated by a slight gap, and tapered at the end. As the nut is tightened and torque is increased, the coned ends of the two halves are forced against a tapered washer. This causes the nut to close tightly around the thread, the slight gap between the two halves allowing for this.

In tests screened on *The Inventors*, a nut made to Pat Williams' design withstood a tension of nearly 27 tonnes. An ordinary nut failed at 17 tonnes, because the wedge effect of the thread tends to expand the nut under tension.

Index

Italics indicate pages with illustrations

atomic absorption spectrophotometer 14, 152*ff*, *154*
Alston, James 47*f*
Aspro 104*ff*
Atmotic ship 22*ff*, *24*
Austin, Herbert 37

Baillieu, W. L. 84*ff*
Bandt, Lewis 14, 124
Barma, P. S. 133
Bavay, Auguste de 83*f*
Beech, Lance-Corporal 107
Benjamin, Lou 14, 120*ff*, *121*
Bland, William 22*ff*, *23*
Boas, I. H. 120
Bottrill, Frank 70*ff*
Brennan, Louis 12, 44*ff*, *45*
Bull, John 9*f*, 11, 16*ff*, *17*

Campbell, Robert 31
cane harvester (Falkiner) 117*ff*, *118*
castors, furniture (Shepherd) 136*ff*, *138*
Chaffey, George 58*ff*, *59*
Chaffey, William 60
Cleve 70
clothes hoist, rotary (Hill) 139*f*, *140*
Clothier, Keith 183*ff*
Cohen, Montague 84
Coolgardie safe 68*f*, *68*
Cotton, Frank 156*f*, *157*
CSIRO 14, 76, 152, 160*ff*, 167, 176*ff*, 180*f*, 183, 192*ff*

Debenham, Mike 187*f*
Delprat, Guillaume 14, 82*ff*, *84*
Dethridge, John 94*f*, *94*
Dungar, Henry 26
Dungar, William 26

Eason, Owen 163*f*
electrical measurement 183*ff*, *184*, *185*
engine, orbital (Sarich) 15, 170*ff*, *172*, *173*
ergometer (Cotton) 156*f*

Falkiner, R. S. 117*ff*
Faraday, Michael 28
Farfor, Jamie 202
flotation process (mineral extraction) 82*ff*, *85*
Flying Doctor Service 114*f*, 116, 135
Flynn, John 114*f*

Gaunt, Ken 144*ff*, *145*
golf clubs, Geometrik 186
Grant, Jack 190*f*, *190*
Gray, George 34*f*
gun, sub-machine (Owen) 12*f*, 126*ff*, *127*, *128*
Gwynne, J. A. E. 32

Hargrave, Lawrence 13, 61*ff*, *62*, *65*
Harrison, James 10, 26*ff*, *27*
Henshaw, David 160*ff*, *160*
Higham, James 10, 32
Hill, Lance 139*f*, *139*

Howard, Cliff 13, 110*ff*, *111*, *112*
Howard, John 34, 36, 37
Hume, Walter 13, 87*ff*, *88*

Interscan 179*ff*, *180*, *181*
inventions, agricultural 9*ff*, 13, 14, 16*ff*, 32*ff*, 38*ff*, 49*ff*, 54, 55*ff*, 58*ff*, 77, 89*ff*, 98*ff*, 110*ff*, 117*ff*, 144*ff*, 167*ff*, 170
inventions, communication 22, 114*ff*
inventions, irrigation 13, 58*ff*, 94*f*
inventions, military 12*f*, 72, 96*f*, 107*ff*, *108*, 126*ff*, 130*ff*, 133*ff*
inventions, mining 14, 82*ff*
inventions, preservation of food 26*ff*, 30*f*, 68*f*
inventions, transport
  air 13, 22*ff*, 46, 61*ff*, 79, 179*ff*, 190*ff*
  land 14, 32, 37, 70*ff*, 79*ff*, 124*ff*, 173, 195*ff*
  rail 45*f*, 158*f*
  sea 12, 22*f*, 26*ff*, 61, 77*ff*
Inventors, The 173, 186, 199*ff*

Jeffrey, H. W. 133*f*
Jordan, John 56, *56*
Julius, Sir George 73*ff*, *74*

Kauper, Harry 114*f*
Kernot, W. C. 44, 45
Kerruish, Bill 167*ff*, *167*
Koerstz, Christian 56*f*, *57*

Lampard, Doug 183*ff*
Lane-Poole, C. E. 120
lawnmower (Victa) 148*ff*, *150*
lift-control system (Port-El) 174*f*, *174*
Ling, Harold 139*f*, *139*
Lyster, F. J. 86

McBain, Sir James 29
McCleary, Everard 110
McKay, Hugh 13, 14, 49*ff*, *50*, 102*f*
McLennan, Sir Ian 187

Mason, Frederick 56*f*
Metcalfe, Ken 141*ff*
Michell, Anthony 12, 77*ff*, *78*
Miller 133
Mills, F. 134
Mitchell, Sir Thomas 69
Mole, Lance de 12, 96*f*
Morris, Augustus 31
Mort, Thomas Sutcliffe 31
Muirhead, A. R. 109

Nicholas, George 104*ff*, *104*
Nicolle, Eugene 30*f*

Owen, Evelyn 12*f*, 126*ff*, *128*, 133

paper (from eucalypts) 14, 120*ff*
Park, Robert 34, 37
pedal wireless 114*ff*, *115*
pipes, spun concrete (Hume) 13, 87*ff*, *88*, *90*, *91*, *92*
plough, stump-jump 9, 11, 38*ff*, *40*, *41*, *42*
Pomeroy, John 130*ff*, *131*
Port, Leo 174*f*, 186, 200, *200*
Potter, Charles 14, 82*ff*
Potter, Edmund 192*ff*, *193*
Presto can 187*ff*, *188*
Pritchard, Ted 195*ff*, *196*

Richardson, Garry 148
Richardson, Mervyn Victor 148*ff*, *149*, *150*
Ridley, John 9*f*, 11, 16*ff*, *17*
Robertson, Tom 134*f*
rotary hoe (Howard's Rotavator) 13, 110*ff*, *112*
Rowcar 163*ff*, *165*, *166*
Russell, P. N. 30

Sarich, Ralph 15, 170*ff*, *171*, 199
Savage, Robert 33*f*
shearing machinery 32*ff*, *35*, *36*
Shepherd, George 136*ff*, *137*

Shmith, Henry 105*f*
Smith, Clarence 11*f*, 38*ff*
Smith, Richard 11*f*, 38*ff*, *39*
Somerville, John 14, 120
steam car 195*ff*, *196*, *197*
stripper-harvester (Bull and Ridley) 9*f*, 11, 16*ff*, *19*, *20*
Sunshine harvester (McKay) 13, 49*ff*, *52*, *53*, 98
Sunshine header (Taylor) 98*ff*, *99*, *101*

tank, military 12, 96*f*, *97*
Taylor, Headlie 54, 98*ff*, *99*
Terry, Burt 163*ff*, *164*
textiles (Repco spinner) 160*ff*, *161*
Thompson, Mel 183*ff*, *184*
torpedo (Brennan) 12, 44*ff*, *46*
totalizator, automatic 73*ff*, *76*
Towns, George 114
Traeger, Alf 114*ff*, *115*
Tritton, Sir William 96

utility truck 14, 124*f*, *124*, *125*

Walsh, Alan 14, 152*ff*, *153*
water purification system 176, *177*, *178*
Weiss, Don 176*ff*
Wild, Paul 180
Williams, Pat 202, *202*
Wilson, James 35*f*
windmills (Alston) 47*f*, *48*
Windsor harvester 167*ff*, *168*
Wolseley, Frederick York 10*f*, 32*ff*, *33*
wool compressor 55*ff*, *56*
Wright, Bob 141*ff*

xerography 141*ff*